LOUSY SEX

Stained
TL
8-30-17

lousy sex

CREATING SELF
IN AN INFECTIOUS WORLD

gerald n. callahan

UNIVERSITY PRESS OF COLORADO

Boulder

© 2013 by Gerald N. Callahan

Published by University Press of Colorado
5589 Arapahoe Avenue, Suite 206C
Boulder, Colorado 80303

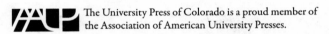 The University Press of Colorado is a proud member of
the Association of American University Presses.

The University Press of Colorado is a cooperative publishing enterprise supported, in part, by Adams State University, Colorado State University, Fort Lewis College, Metropolitan State University of Denver, Regis University, University of Colorado, University of Northern Colorado, Utah State University, and Western State Colorado University.

♾ This paper meets the requirements of the ANSI/NISO Z39.48-1992 (Permanence of Paper).

Library of Congress Cataloging-in-Publication Data

Callahan, Gerald N., 1946–
  Lousy sex / Gerald N. Callahan.
     pages cm
  Includes bibliographical references.
  ISBN 978-1-60732-232-0 (pbk.) — ISBN 978-1-60732-233-7 (ebook)
  1. Self. 2. Identity (Psychology) 3. Biological psychiatry. 4. Psychoanalysis. I. Title.
  RC489.S43C35 2013
  613—dc23
                          2013007245

Design by Daniel Pratt

22   21   20   19   18   17   16   15   14   13   10   9   8   7   6   5   4   3   2   1

Contents

Acknowledgments

A different version of "First Self" appeared in *Emerging Infectious Diseases* 1 (2005). Different versions of "Layers of Self," "Dreams of the Blind," and "The Mysterious Visions of Jean-Baptiste Pierre Antoine de Monet, Chevalier de Lamarck" appeared in *turnrow* magazine under the titles, respectively, "Just the Two of Us" (2[2] [2003]), "Blindsight" (6[1] [2009]), and "Darwin's Dream" (4[1] [2005]). Other versions of "Self in the Soil" appeared in *Emerging Infectious Diseases* (9 [2003]); and *Faith, Madness, and Spontaneous Human Combustion: What Immunology Can Teach Us about Self-Perception* (New York: St. Martin's, 2003). Portions of "Gathering Our Selves" originally appeared in *turnrow* (Vol. 5[1] [2007]); *Science and Spirit* (14 [2002]: 60); and *Infection: The Uninvited Universe* (New York: St. Martin's, 2006). And another version of "The Wizards of I" ("A Dog's Life") appeared in *Many Mountains Moving* (Winter 2007).

Prologue: Leonardo's Dream

He certainly hadn't planned to spend his morning wandering around looking for inspirations, as though they might be startled from the shadows like mice. He kicked at a stone and sent it flying across the Via Mercanti. As he walked into Milano's great piazza, pewter-colored clouds rolled in from the mountains to the north and blocked the sun. The landscape fell into a flat brown lake. Leonardo kicked at another stone, missed, and stubbed his toe against the red cobbles. He cursed and hobbled to one of the benches ringing the piazza. Sitting, he lifted his injured foot onto his knee. The wind, mocking him, played in his wild hair. Leonardo cursed again as he reached to push the hair from his face. He stared solemnly at his toe.

For years, this man had studied human anatomy, patiently slicing through skin and muscle, bone and gristle, splitting open eyeballs and undressing the dead in the most intimate of ways. On top of that, he was a meticulous observer. As he worked, nothing escaped his attention, and he laid each of his dissections carefully onto the pages of his notebooks—every tendon, every fascia, each nerve found its way there. He knew human bodies as well as any other living, or likely dead, man.

Just now, none of that helped. The pain in his toe had emptied his mind. The wind from the mountains began working its way into his bones. The piazza smelled of burned oil. Forty-one years old and, on this particular day, nothing to show for it. This was too much to bear. Like the birds scavenging among the cobbles, ideas always found *him*. His brows bristling with intent, he scanned the world before him.

The spring weather, the smell of fish in the markets, and the solid feel of his feet on the cobbles of the street reminded him life was short. He didn't have time for lapses like this. He continued to rub his foot.

Across from where he sat, a young couple had found their way into a darkened alcove of the partially constructed cathedral. Thinking no one was watching, they fell into an amorous embrace and kissed one another deeply. Hands moved about, cloth rustled, the woman's face flushed red.

As Leonardo watched, their embraces grew even bolder and more intimate: a hand dropped between them, a fastening was pushed aside, another.

To his surprise, Leonardo felt a rising in his own loins and a curious sort of embarrassment at his reactions. But he couldn't stop himself from watching the two lovers as their hands found still more heat and their caresses deepened. Finally, as the two bodies collapsed into one, he knew what he would do.

A thing no one had ever attempted before. As the thought uncoiled inside his mind, he smiled, then lowered his injured foot onto the stones of the street and rose from his bench. Like startled birds, the lovers cried out and fluttered, reaching to cover themselves. But now, for Leonardo, the two might as well have been in Genoa. Leonardo's mind had taken flight, and even as the wind worked again at his beard and long, thin hair, he continued to smile. With each step, he watched the slow flex and relaxation of the muscles in his thighs.

The nearer he came to his studio, the more excitement he felt. Today, he would begin his most intimate of drawings. It would be unlike anything anyone had done before. He climbed the stairs two at a time.

Inside, amid the smells of preservatives and plaster and brine, Leonardo pushed his other projects from his drawing table, allowing them to drop to the floor. On a clean piece of paper he began to sketch—a man cut in half lengthwise. Once he had the man's outline in place, Leonardo's hand pulled out the shape of a heavy-breasted woman, and just as he had seen the lovers on the piazza do, he drew the two together at their loins.

Spines and hearts, bones and brains fell onto the page. Muscles took shape and backs stiffened. The last part he knew would be the most difficult, the part where the two became one. He had seen all of the rest, more times than he could recall. But this, the most precious of touches, he had only felt.

That, he thought, *will be the most difficult.* Penises he could draw, he'd seen dozens, even in cross-section. Vaginas the same. But he had never seen this moment like an anatomist, from the inside. *It would come, though. He was certain. When at last he reached for it, the image would come.*

His left hand fell once again upon the page, his right idly rubbed his thigh. Before him appeared a sight no man or woman had ever seen. The art of it, the science of it, all lay there on the page—redolent. He stood back and smiled at what lay before him. He titled his creation *The Copulation.*

The morning after, he invited others to see what he had done. This is what he said to them: "I expose to men the origin of their first, and perhaps second, reason for existing."

His drawing offered the consummation of a romance begun centuries, perhaps millennia before—the wedding of art and science symbolized as an intricate union of anatomy, sexuality, and human ardor. A marriage made not in heaven but in the fertile minds of men and women. Inside this one sketch was the ultimate union of two selves, Leonardo's own prevarications, sex, art, the roots of Darwin and Lamarck, hope, and lust. It was all there, all at once.

And, like mantises, the mates devoured one another. Now there was only one thing, androgynous and seemingly immortal. And for a while that thing flourished. If ever there had been any question about this union, about whether science and art were made for one another or by one another, it fell silent that day.

What Leonardo had done was to marry art and science as no cyclopean scientist or artist could have done. *The Copulation* was more than art or science, it was a thing of its own, at once holding the stories of science and the humanity of art. And, of course, it held Leonardo as well.

In 1540 or thereabouts, in apparent homage to Leonardo, Andreas Vesalius, a Flemish anatomist, went so far as to hire a student of Titian's, Jan Stephan van

Calcar, to illustrate the *De humani corporis fabrica*—Vesalius's studies of human anatomy. That book's fusion of anatomy and art would dominate human expectations and perceptions of anatomy for centuries to come. Art saturated with the ooze of science. It was intoxicating. For nearly 200 years, the lovers fed upon one another.

But in every marriage, the fruit of certainty nourishes a seed of doubt. It may take decades, but in ways no one ever foresees, often that seed sprouts from the soil and devours what it finds.

Science changed.

A mere 100 years later, in place of Leonardo's ample breasts and *amo ergo sum* was Descartes's y = mx + b and *cogito ergo sum*. The rift widened. Art turned to saintly food for sustenance. Science fattened itself on reason. Abruptly, the soul became an amorphous spiritual thing belonging to Rome; the body and the brain became the playthings of science. Art from the soul, science from the mind. Vows were broken, lies told. Everything changed.

Ignorant of all this, about 500 years later, I sat in the sun one afternoon and read an essay about wood lice. The title of the essay was something like "The Exception That Proofs the Rule." As explained in the essay, the common expression had not begun as "the exception that proves the rule"—which makes no sense. It began as "the exception that proofs the rule," meaning an apparent exception that tests (proofs) the validity of a rule. And this author's exception was wood lice. The rule being proofed was the general assumption that sex ratios among sexually reproducing species will always be about 50:50, as it is in humans. The belief being, I guess, that such a ratio is inevitable because it provides an optimum number of opportunities for mating to the maximum numbers of males and females.

It turns out that sex ratios in wood lice are nowhere near 50:50. In all species of wood lice, females vastly outnumber males. And in some species of wood lice, there are no males at all. The author of my essay spent a lot of time studying one species of wood louse (I don't recall which one) trying to figure out just what was going on here. Previous students of this species of louse had failed to find any males. To the author that seemed improbable. So he began a careful investigation of these lice.

Wood lice are isopods—crustaceans with seven pairs of legs. Most everyone has seen a wood louse—roly-poly bug, sow bug, potato bug—about a half inch long, armor-plated, crawling out from under rocks or in the leaf litter under the deck. When poked at, wood lice roll up into tight little chitin-covered balls.

Careful study revealed that in this particular species, males did occasionally appear, but rapidly died, apparently without ever engaging in sexual reproduction. Which seemed pretty weird.

Eventually, the author found that a collection of eggs appeared inside all female wood lice as they matured. While still inside the louse, these eggs hatched, and invariably only a single newly hatched louse was a male. That male immediately impregnated the now-hatched females, who then began to eat their way out of their mother, each already pregnant. The male, having served his purpose, was dumped into this world, staggered about for a bit, and died.

The author concluded that 50:50 sex ratios didn't apply to species that carefully protected their males and allowed them to mate with many females. And that, in spite of the decreased opportunity for genetic diversity, somehow made perfect genetic sense.

The author was fooled by the scientific appearance of things. To the lice, he brought only the science half of the androgyne. The artless half concerned itself only with one thing at a time—an isopod. Because of that, the author missed how the simple art of the wood louse would eventually change the way we all think about not just bugs but ourselves as well. The whole lousy story about isopods (and many other pods) appears in the chapter titled "Lousy Sex."

But formulae like y = mx + b and words like *cogito ergo sum* and all the others that followed didn't end the androgyny. They only disfigured it, forced it to show one scarred face at a time.

The mantises still clasp one another, and in the darkness of those multifaceted eyes there is still a hall of mirrors. The stories in this book explore those reflections, consider the personal and artful implications of the deep insights of biomedical science. From the still-life meanings of dementia to infectious sex to the flame of immortality, the book spans the imaginary gulf between the intimate and the scientific. It is a book full of stories that feed on the heat created by the essential intercourse between science and art.

Each of the tales that follows considers the art and science of what it means to be a human being in the same way a painter might explore how a weave of colors can change how we experience the sea or an old woman. Artists have changed forever the way I will think about the night sky. So have astronomers and physicists. But for me, the greatest insights about myself and my universe came when I stitched together Jocelyn Bell's pulsars and Vincent van Gogh's *Starry Night*, or when I interwove Josiah Willard Gibbs's free energy, chemical thermodynamics, and Rainer Maria Rilke's panther. In these spaces are stories about where we came from and who we are, about why we care and where we will end. This book explores those spaces and some of those stories.

LOUSY SEX

ORIGINS

Where "I" Comes From

Like the fossilized tooth of megazostrodon, beneath the enameled surface of the word "I" lies one of the great stories of our past—the origins of selves. Surely there was a time without selves. How and why did "I" evolve from not-"I"? What did the first "I" look like? Like the lizard brain, underneath the layers of more sophisticated and more civilized "I"s, first "I" is still there. Underneath everything else, beneath all that we call human, among the rubble left when the lizard brain finally rules, it is there. There, my mother and I achieved purity.

Then there is the protective "I," the one that each of us carries around each day. Where, among all of the blood and sinew and gristle, does everyday "I" live? Where are our selves?

The answer, of course, is everywhere.

Surprisingly, immunologists spend more time thinking about the word "self" than most of the rest of us. Though often overlooked, every one of us has a living system whose sole function is to separate self from nonself—an immune system. Our immunological sense of self is all that stands between us and every other living thing, the one sense we cannot live without. That sense of self is born inside human thymuses. That isn't where most of us would choose to look for a sense of self. Most of us would look instead inside of human brains for human selves. And, of course, you need look no further than Phineas Gage and his tamping rod to realize that changing human brains does alter human selves. But that takes nothing away from the immune self. Perhaps two "I"s inhabit each of us.

One human disease provides a clue. Multiple sclerosis is a war between human immune systems and human nervous systems, a war between the selves. Three of my friends have or had multiple sclerosis. In these women I have seen another sort of self. If we wish to understand "I," we need to know about human immunity, the jackets that surround human neurons, layers of self, and the war between the selves that we call multiple sclerosis.

———————

Are "I"s born or, like memories, do we gather "I"s from the world around us and add to or subtract from them every day? And what are the smallest pieces of "I"s we cannot be "I"s without? For most of the first year of a baby's life, it cannot distinguish itself from its mother. For that year, "we" is "I." Clearly, at the outset, we are more, and less, than "I." Constructing an "I" takes a lifetime. We build those "I"s from bits and pieces of the people and the world around us.

Through an immunologist's eye, "I"-building looks a lot like a cross between sculpting mud pies and making powdered donuts. Mud, dirt, soil, grime, and dust hold one of the keys to selfhood. It is literally true that we are dirt and unto dirt we shall return. But that is not just because the soil feeds the plants and animals that feed us and that we will one day feed in return. Nor is it because some god cursed us with mortality. Soil holds bacteria and we cannot construct an "I" without bacteria.

No organism on Earth has had more biological success than bacteria. Bacteria outstrip every other living thing in numbers, in mass, in distribution, and in variety. And, because they came first and because there are so many of them in so many places, bacteria have extorted an agreement from all the rest of us. That agreement reads something like, *Live with bacteria or don't live at all.* That agreement binds all of us and, in this world, we must live as bacterial symbionts—sharing our lives with unimaginable numbers of these creatures.

The whole thing begins when our mothers intentionally infect us during labor. And it ends only when flame or prey consume the last of us. In truth, we literally gather our selves from the world around us. We extract bits of "I" from the dirt we eat, from the people and the pets we touch, from lovers and from the food we eat, from door knobs and church pews, from computer keys and borrowed books.

And those infections change each of us, make us who we are—the pointillist "I." The long story of the human genome lays it all out; humans would not be humans without our infections. Ultimately, the search for self-knowledge leads to our infections, past and present.

Cave painting, Patagonia, Argentina

I

First Self

> If selves are anything at all, then they exist. Now there
> are selves. There was a time, millions (or billions) of years ago,
> when there was none—at least none on this planet. So there has to
> be—as a matter of logic—a true story to be told about how there
> came to be creatures with selves.
>
> —*Daniel C. Dennett,* "The Origins of Selves"

Slowly, purposefully, my mother unbuttons her blouse. The blouse is blue with small white flowers, and the tail is tucked firmly into the elastic waistband of her salmon pink pants. Beginning at the top and moving down, she works carefully at each of the small plastic buttons.

"Mother," I plead with her, "you don't need to do that."

She smiles at me and continues unfastening buttons. Beneath her blue blouse, her padded cotton and elastic bra begins to appear. Her breasts swell pallidly above the brassiere.

The room is not well lit. The curtains are drawn, as they always are, against the sun. But I can see more of my mother than I wish to. My

DOI: 10.5876/9781607322337.c01

father, sitting here with me, says nothing. My wife, Gina, and the two other women in the room also sit silently as my mother undresses herself.

I smell my mother's perfume as she works at her blouse, her perfume and the lotion she slathers herself with every morning. I see the wrinkles beneath her arms, the flaps of skin at her elbows.

She pulls off the blouse and stops, standing before us all, the blouse hanging in her right hand. "Isn't it pretty?" she says and turns slightly so each of us can see her brassiere. My mother is eighty-two years old and mostly naked to the waist in front of her son, her husband, her daughter-in-law, and two perfect strangers. Her gray hair sprays from her head in nearly every direction. Her back is littered with small brown moles. And her dark eyes, fallen far back in the sockets of her skull, flutter from face to face.

"Do you need a brassiere?" she asks, grinning thoughtlessly at my father. His breasts, sagging with his eighty-six years of life, nest like doves beneath his gray knit shirt. He lowers his eyes and looks away from her.

This is not, of course, my mother. My mother would never have bared this much of herself in front of strangers, and certainly never in front of her son. My mother was quiet, shy, prudent.

And this *is*, of course, my mother. Nearly naked in front of me and strangers. Proud of her new bra, badgering my father. I know it is my mother, because it is clearly her face, her hands, her dried-out, fungus-ruined feet here before me. But things have changed.

The nondescript, nappy, brown carpet is just as it always has been. The counters are still lined with the detritus of a middle-class life. The cheap fan still chops at the air overhead, and the plastic draining board—with its plastic dish rack—still drips dishwater into the same stainless steel sink. The clock, with its three golden globes, rolls on the ball bearings of the hours, just as it always has. But my mother is mad. And the five of us have gathered here today to evaluate her for custodial care. Custodial care—that sounds as though we might turn her over to the janitors at the university where I work. As though they might know what to do with her since we don't. Repugnant or not, though, we have no more time to twist our tales.

THE MEANINGS OF SELF

Among other things, I am immunologist. I have spent my life studying the intricate paths by which we protect our selves from this infectious world. Self, nonself, and why the two should never meet. But as a son watching his mother disintegrate, I am cut adrift.

My mother's self, the thing that was her for all these years, the thing I had imagined fixed as flint beneath her bones, has fractured. Shattered like a crystal vase on concrete. It is one thing to watch feathers grow from chicken-skin grafts on self-confused nude mice. It is quite another thing to watch your mother undress herself in front of total strangers.

————

Merriam-Webster says self is "the entirety of an individual, the realization or embodiment of an abstraction." The realization or embodiment of an abstraction. I don't know what that means. It seems woefully incomplete and miserably metaphysical. As though only abstractions will do to speak of my mother's disappearance—no matter how concretely and obviously she is disappearing.

Beyond *Merriam*, Sir Frank Macfarlane Burnett, considering the same issues, chose a laboratory and microscope rather than the caverns of the mind for his explorations. Through the lenses of that microscope, Burnett watched as an ameba devoured and digested another microorganism.

> The fact that one is digested, and the other not, demands that in some way or other the living substance of the ameba can distinguish between the chemical structure characteristics of "self" and any sufficiently different chemical structure as "non-self."

Biological self self-evident in the simple battle between devourer and devoured.

And here's Burnett later, in contemplation of an immune response:

> The failure of antibody production against autologous [self] cells demands the postulation of an active ability of the retciulo-endothelial cells [immune system] to recognize "self" patterns from "non-self" patterns in organic material taken into their substance.

Biological self, not as concept or thought, not as abstraction, but as the solid fist of antibodies as they take hold of infecting germs but leave the body itself untouched.

In other words, even the most primitive of us don't regularly eat ourselves. And even the most complex among us don't regularly mistake our bodies for infectious enemies and destroy the very thing that sustains us. It is the unique way of our being that our digestive and immune systems ignore us while they chew away at the rest of the organic world.

Something substantial, a thing apparently very different from Webster's self.

The fact that, on the surface, these two selves—the abstraction of *Merriam-Webster* and the rock-solid self of Frank Macfarlane Burnett—seem incommensurate, we probably owe to René Descartes.

WHERE MICRO- AND MACROORGANISMS SEPARATED: THE DIVIDED SELF

Descartes, a mathematician and philosopher, found himself one day deeply concerned with the reality of things. What could he truly trust? What was rationally and irrefutably real? We all know that we make mistakes at times about what is real—the monster under the bed, the shadow in the closet, mirages, purpose. Most of us just shrug it off, but Descartes was not so easily mollified. He wanted to know for certain. So Descartes eschewed the laboratory (he missed the microscope by about twenty years) and secluded himself in a darkened room at the back of his chateau. There, he set out to discover what was demonstrably real, trustworthy, *certain*. What he new for sure about the world and his self.

First, Descartes considered the things we learn through our senses—the stuff we see, hear, taste, touch, and smell—the physical world that apparently surrounds us. Is any of this truly real, unquestionably real? No. Almost immediately, Descartes realized our senses regularly fool us. Dreams provide hard evidence of that. While we are in a dream, we become completely absorbed with a false reality. Dreams do not announce to us that they are not "real." And many things in the so-called real world do not announce to us that they are false, things like an obviously flat Earth, optical illusions, and sleights of hand.

The reality we come to know by touch or sight or smell or sound or taste may be as ghostly as our dreams. All of the natural world became questionable. Descartes's vision narrowed.

Since Descartes was the inventor of analytical geometry, he turned to the reality of mathematics—a priori knowledge, knowledge accessible without sensory perception. Because of Descartes's deep investment in mathematics, he thought a priori knowledge to be a thing of stone, beyond reproach, above suspicion. But as he delved deeper, he realized that some "evil genius" might have fooled us all about the reality of mathematics. Evil indeed. Mathematics might be nothing more than an elaborate ruse and have nothing whatsoever to do with "reality" (as many of us suspected even in grade school). In the end, Descartes was forced to abandon mathematics and all a priori knowledge as well as sensory knowledge.

Without the physical world, without mathematics, the only things left to Descartes were his own thoughts. He realized, at the last, that rationally and philosophically he could not question the reality of the questioner. It simply wouldn't make sense. So his questions proved his own existence, even if he could not establish the existence of anything else. *Cogito ergo sum.*

If Descartes had been a microbiologist or immunologist like Burnett, things might have ended differently. But for the mathematician, the world devolved to one man's thoughts. Descartes rested, then, in the midst of an absolutely solitary universe.

According to him, two types of things existed: the seemingly real but demonstrably untrustworthy things of the physical world (*res extensa*), and the only truly real things, things of the mind (*res cogitans*). These were two completely separate kinds of things. The world outside our heads was full of machines and ghosts, including our own bodies. Our thoughts were concrete, real, essential; our bodies and our worlds abstractions. Descartes had scalpeled the self off the body.

The self, he claimed, was something other than the stuff of the physical world that surrounds us. Selves did not come from the same stuff that trees and stones and arms and legs and knuckles and immune systems came from. Selves came from somewhere else. Self-stuff and body-stuff were distinct and immiscible.

But almost 400 years later, as I watch my mother fumble with the tails on her blouse, I recognize a certain absurdity in the perspectives of René Descartes.

THE REALITY OF BIOLOGICAL SELF

We finally convince my mother to put her blouse back on and button it. It takes her two tries, but she now has each button in its proper hole. I am embarrassed. My mother seems unabashedly pleased with herself, what remains of it.

She smiles again at me. Sitting, now, quietly while everyone else tries to deal with the shadows she has cast across the room. My father still sits with his eyes lowered, waiting for someone to tell him what to do. I turn to one of the women seated across the room. I look for some sort of forgiveness, some sort of reassurance that my mother's antics haven't ruined all of this for all of us.

"She will do perfectly, I think," Jennifer says.

"Does she wander at night?" Melissa asks.

Our selves are not something ethereal, something forged from the stuff of a separate reality. Our selves are no different from our livers or our hearts. Our selves are just as susceptible to the effects of breeding and infection and aging as any other part of us.

So, just as there is a biology of reproduction or of respiration, there must be a biology of self. Who we are is not simply a matter of spirit or story or thinking. It is in our genes—those we are born with and those we acquire.

That's important, because genes arose and were preserved over eons to protect us, to provide each of us with some specific edge in the struggle for survival and reproduction. The genes we have come from a very long line of survivors and reproducers. In among those genes is the template for self.

"She does wander at night," I add, wishing I didn't have to. "Twice, that I know of, Dad found her outside the house in Kanab, Utah, making her way down the road toward town." The first time she wasn't even wearing the bottoms of her pajamas. She was naked from the waist down.

When he stopped her and asked where she was going, she said, "Home. I'm going home." My father, though he tried, could not make her understand that she was home.

"When would you like to move your wife?" Melissa asks my father. I watch to see his response.

The Origins of Infectious Disease and the Reason for Selves

If selves are born inside of genes, then, just as with all things biological, there must be an evolutionary advantage to selves and an evolutionary history of selves.

In the beginning there was RNA (probably)—ribonucleic acid, strange curlicues of chemical bases that snapped together spontaneously in the witches' brew that was the primitive seas. Then there was DNA (deoxyribonucleic acid) that twisted itself into long chains and then wrapped itself up in fatty acids. Later, true cells appeared. Life—bacteria, archaea, prokaryotes, protists, eukaryotes. Everything was suddenly possible.

But only if everyone followed three basic rules: (1) eat; (2) don't get eaten; and (3) reproduce as quickly and as often as possible. Three rules alone that would account for all who followed. But at least the first of these rules required a certain insight, a certain pair of interlocking concepts—self and nonself. Nonself is food; self is not.

First-self had walked onto the stage. Pronouns became meaningful. A simple "it" was no longer sufficient to describe everything. "Me" and "you" were necessary now. While sense of self was, perhaps, a long ways off, self was there, that day, swimming in a thin broth of "other."

The next major step up from the muck was bacteria or archaea, or something much like them. These were, after all, living, respiring, and reproducing creatures. But bacteria suffered from one huge drawback—each of them had only one cell to work with. That meant then, and still means today, that most bacterial cells had to do everything, all of the time, all at once. Each cell had to see, hear, touch, taste, and smell. Each cell had to eat and excrete, reproduce and think. Each cell had to make everything that was needed for the survival of the individual. Because of this, bacteria—though remarkable survivors—weren't and aren't much good at anything beyond simple survival—bacterial poetry, for example, is unreadable.

One day, all of this began to change. A few cells got together cemented themselves to one another with some new glue. A protoplasmic hand reached into the void and another hand took hold. The door of opportunity swung wide open. For the first time, individual cells were freed from the slavery of necessity. No longer did anyone have to be everything for everyone. No longer did anyone face everything alone.

Cellular specialization took the world by storm. Some cells stopped eating and became eyes (or something that would one day become eyes), others ears, others nerves, others muscles—there were no limits. Taste buds, antennae, pincers, intestines, hearts, tails, legs, arms, muscles, bones, livers, lungs, hair, nails, claws, blood, hide, and horn were all within reach.

But almost immediately, everyone saw that cellular specialization alone led nowhere. Before the gift of multicellularity could be had, before grandeur, they needed selfishness. The first few of these complex multicellular creatures probably shared everything with everyone. After all, they had no means to distinguish among themselves. All that I have is yours, not because of altruism, but because I cannot tell you from me. Such largesse defeated the whole purpose of cellular specialization. What benefit is there to eyes, if what I see I share with the blind who surround me? Remember, there are rules. *I* come first. *I* am not to be eaten by others. *I* am to eat others. *I* am to reproduce first. The things *I* see are for *me* and for me alone.

But up until now, "I" had meaning only in terms of food, only in terms of the external world. "I" don't eat "I," "I" eat "not-I." "I" eat what is out there, not what is in here. Something more was needed. At a higher level, what or who was I and what or who was not? I could not decide. Before I could reach for the stars, I had to reach within and find some way to know myself from all the others.

Without a sense of self we are less than bacteria.

If I am to keep what I have earned, if I, and I alone, am to benefit from my mutations and absorptions, my specializations, my senses, my motility, then I must know self from not-self. My eyes must be for my self. My thoughts must be my own. My heart must beat only for me. I must keep all that I can to my self at the expense of not-self or I have gained nothing.

Selves leave no fossils. So we cannot know for certain how or when the first colonial (multicellular) organisms came to sense their selves. But *biologically, biochemically, basically,* they had to know, everything depended on it. And the biology and the chemistry of that knowledge were and are the only things to keep us apart, to prevent us all from slipping back into the singularity of a living ooze known only as it.

The evolved self. The self geneticized. A protein marker, perhaps, carried by every cell inside of every one of us. A passport to be checked and rechecked at every interaction. A self to be validated over and over as coded instructions, warnings, and alarms passed from cell to cell.

For a few millennia, that was probably good enough. But life was changing. Somewhere along the line, microorganisms discovered the miracle of parasitism. Once inside another's membrane, food cost nothing, life was simple, and reproduction was almost guaranteed. There was nothing else quite like it— "Money for nothin' and your chicks for free."

Self-discrimination was no longer enough. Once others learned to hide inside of self, force was needed to maintain boundaries. Now we needed immunity, a fierceness to keep us whole.

Once again, if our own mutations and adaptations were to serve us, the integrity of self was essential. Infectious diseases posed the first great challenge to the biological preeminence of self. Immune systems arose and quickly found intricate ways to detect and terrible means for destroying nonself.

In the beginning, biological self was probably nothing more than a simple system for recognition of other as food. Now the self had teeth. Now the self rose like a shield to stand between us and those who would destroy us.

INFECTION, IMMUNITY, AND THE WEAVE OF SELF

Over time, the self grew. Like the brain itself, layers upon layers of self formed inside of living things. Later, much later, like the cerebral cortex, there arose psychological self—self-conception, self-perception, self-deception—something to be explored in the next chapter. But still, like the amygdala in the brain, beneath the complex and sophisticated self there beat the heart of a beast focused only on food, survival, and sex.

But unlike the brain, the layers of biological self were strewn throughout the body. In between the layers was immunity and infection. And those, it seems, tied together the layers themselves and stitched biological self to psychological self.

The proof, as they say, is in the pudding, or in this case in the custard—the gray and white custard we call a human brain.

Important exams, other men and women, public speaking, air travel—things that frighten or threaten cause our immune systems to stumble. Perceived dangers cause a stir in the hypothalamus—a small organ at the base of the brain that controls a host of human functions. In response, among other things, the hypothalamus secretes corticotropin-releasing hormone (CRH). CRH stimulates the pituitary gland to release adrenocorticotropic hormone (ACTH). ACTH induces the adrenal glands to produce cortisol. Cortisol suppresses T lymphocytes (an essential white blood cell in most types of immune responses). A human thought changes the lives of lymphocytes—the selves synergized. Two become one.

Interestingly, other sorts of threats—particularly infectious ones—enhance other parts of the immune system, especially inflammation. Simply looking at pictures of people who appear to be ill is enough to cause men and women to produce more of a substance called interleukin-6 (IL-6). Interleukin-6 causes a whole series of other things to happen, especially inflammation—something that could be very handy in someone about to be exposed to an infection. Interleukin–6 also changes human thinking. Men and women with higher levels of IL-6 lose interest in sex, in eating, in exercising, and in social interaction in general—also potentially handy behavior changes in the presence of others with infectious diseases. All of that happens because of the effects of IL-6 on the human hypothalamus. Reverse synergy.

Cytomegalovirus and *T. gondii* have been implicated in the etiology of schizophrenia, and individuals with bipolar disorder are more frequently infected with herpesvirus type 1 than those without bipolar disorder. Mice born to mothers infected with influenza virus never develop a lust for exploration. And infections by any of a variety of other parasites, bacteria, and viruses also change animal behavior in unexpected and consequential ways. Infections confuse us.

Infections may also destroy us. In a study conducted on more than 3 million U.S. military personnel from 1988 to 2000, the strongest predictor for developing multiple sclerosis was serum antibodies to Epstein-Barr virus (EBV)—meaning that people infected with EBV were more likely to develop multiple sclerosis than people never infected by the virus. A finding that pointed a wicked finger at EBV as a possible cause of multiple sclerosis. More about that later.

Viral or bacterial infections also seem to play a role in several autoimmune diseases, like rheumatoid arthritis, diabetes mellitus type 1, Crohn's disease, and some types of thyroiditis.

Maybe some infections smudge the immunological boundary between self and not. And it seems likely that as our understanding of the relationship between infection and immunological self-perception deepens, there will be more places found where the boundaries between the immunological and the psychological will crumble.

Also, without infection, animals never develop any functional sense of immunological self. Human autoimmune diseases, such as type 1 diabetes, show a significant correlation with the development of schizophrenia and other behavioral disorders. Cause-and-effect relationships remain obscure, but these findings suggest a link between infectious disease and our immunological as well as psychological perceptions of self.

The gossamer web that it weaves remains mostly hidden, but the hard reality of self is self-apparent.

LAST SELF

All her life my mother preferred things simple. She liked jam better than jelly. She loved cornbread and blackstrap molasses, white gravy, melons, "Amazing Grace." She never cared much for driving. She grew up poor, truly poor. Maybe poverty burned up all the fuel she was saving for complexity before she ever found any. Regardless, her tastes never changed. She was always most comfortable with ordinary things.

I remember that about her, her simplicity. But by the start of the second year of her custodial existence, no matter how hard I try, I can no longer remember much of anything else about who she once was. I can't recall when her hair might have been brown and combed, her pants not fat with diapers, her smile less vacant.

The crater of her face is gaunt and empty now. Every cold and every flu that comes along wracks her lungs and sends mucus cascading from her nose, across her mouth, and onto to the knit sweaters she wears against the cold. Places where her self had played across the geography of her skin are abandoned now. Empty lots overgrown with weeds.

With me, hers is the purest and truest indifference. I love her for that. I sit by her for hours simply to bathe in the low, warm light of that indifference.

The smell of urine is everywhere. We are without pretense.

She speaks her stories aloud; I act as though I'm listening. Over and over, she spins those stories around me as though they might protect me from something I will have to face when she is gone. In the end, as her self flees, it takes her stories as well.

Layer after layer of her peels away until all that is left is the bare wood of her. The weathered boards scoured of paint by the raging storm. Underneath, there is fear, often; a fervent sexuality, always; and hunger. The embers of a fire that first flickered billions of years ago. Life, infection, and her own defenses have stripped her of everything else.

The last day she speaks to me, she is dressed in red sweatpants and a loose purple jersey top with long sleeves. Her nose is running. When I come into her room, she was is lying in bed staring at the ceiling. She rarely speaks complete sentences anymore. She never recognizes me. I sit next to her and for several moments say nothing.

The room is split in two by a red blanket hung as a curtain. On the other side of the curtain, there is another bed. Sometimes Mom has roommates, but for the last week or two, no one has occupied the other bed. Full or empty, the other bed holds no interest for Mother. The floor is covered in spattered beige tiles.

I enjoy these moments, sitting quietly next to my mother—moments stolen from reality, shielded from uncertainty. I am not an immunologist here, just an old woman's son wishing for things that could never be. I reach out and hold her hand, thin now, with swollen blue veins and knuckles fat from inflammation. She turns to me.

"Hello," I say.

"Hello," she says, with obvious pleasure. Then she lifts my hand to her lips and kisses my fingertips.

"How are you?" I ask because whether someone is dying or not, whether someone is bleeding to death from a severed limb or someone has just finished a pastrami sandwich, that question inexplicably comes to my lips first.

"Fine," she says and turns back to the ceiling, smiling. "I'm fine."

I wipe her nose.

This day, she wears no lipstick, and the aides have taken her bridge from her mouth. Four of her lower front teeth are missing. Her tongue falls through the opening when she speaks and twists her words.

"What would you like to do today, Mom?" I ask, not really expecting anything.

She looks up at me with eyes deep brown as mahogany. She purses her lips beneath her small mustache. And for a moment her eyes move off to one side as though she is actually thinking about what I've asked. Finally she looks back into my eyes and says to me: "I'm hungry."

"Then let's eat."

I lift her into her wheelchair and roll my mother down the tiled hall to the dining room. She eats Salisbury steak and mashed potatoes. She eats green beans and corn. She eats peach pie with ice cream. And likely she would eat even more if anyone offered it.

As she eats, she stares across the top of her fork at the brown plastic tabletop. I watch her chin moving with her food and her eyes as they wrestle slowly with nothing. Neither of us speaks while she eats. There is really no need.

Matryoshka: A set of seven Russian nesting dolls

2

Layers of Self

Although its biological separation is immediate, it takes
nearly a full year before a human infant realizes it is an entity distinct from its mother—a self-aware creature unto itself. Curiously,
the final immunological separation from the mother takes place at
about the same time, and then the real battle begins.

SANDY

"Unfair," she says. "That's how it feels." Her words ring with certainty.

I wasn't expecting that. Sandy and I have been friends for years, but
we have never spoken like this before. Sitting in the small gray cubicle
where she works, her words fall like stones—solid, cold.

We're together this morning because I asked her to talk with me
about her MS—multiple sclerosis—a situation quite different from any
other I know. Because she and her disease are so remarkable, I am trying
to write a story about both, and about me.

Outside it's July. Trees full of leaves, freshly mown lawns covered
with children, pavement hot as griddles. Inside it is cool and quiet. I

DOI: 10.5876/9781607322337.c02

wasn't expecting "unfair," and for a moment I can't decide where to go. The silence grows prickly.

"Has it changed you?" I ask finally, unable to think of anything more specific.

"Yes. For one thing, I've decided I won't have children," she says, shrugging off my stupidity. "It just isn't worth the risk, no matter how small, that I might pass it on to my son or my daughter. I always wanted children. But not now."

What a shame, I think, that a woman as beautiful as Sandy will never have children.

"Do people treat you differently?" I ask.

"Last week, one faculty member accused me of being drunk at 9:00 in the morning. It's true I can't walk a straight line anymore and sometimes I slur my words. But drunk? I want people to know that it's the disease, not me. I can't help it."

Sandy is thirty-five years old, petite, brunette, with a face like a cameo, easy to like, and a smile that flickers as quickly and as brightly as flashbulbs. Seven years ago the doctors told her that her immune system had begun a relentless attack against her brain and spinal cord. They said it was multiple sclerosis, this disease. The name seems a bit sterile to me, too easy to say out loud. I think those who named it didn't give the name much thought.

"When I was first diagnosed, the people I worked with worried that they might catch MS from me, that it was infectious. That was hard. And though I try, sometimes I can't help but believe that this is punishment for something I did." For the first time she appears near tears.

"I believe, though, that everything happens for a purpose," Sandy continues, her resolution glistening in her serious eyes, her smile again flickering at the corners of her lips. "I may not understand it now. I may never understand it. But I believe it."

And it's completely clear that she does.

Punishment?

"When I'm done with this, do you want to read it?" I ask her.

"I don't read as easily as I used to."

"I could read it to you."

"No, I'll have my S.O. read it to me. He'd like to read it, too."

It takes me a while to get "significant other" out of that and Sandy's mischievous grin.

The tingling, the slurred speech, the angular gait, the accusations of drunkenness, the worried stares of the unaffected, and the revolt of her own body and mind arriving each day as wakefulness claims her.

Unfair. Perhaps.

But it isn't like Sandy did nothing to deserve her disease, is it? Even Sandy knows better than that. Sandy is, after all, of German descent. And she was born a woman. And she did grow up in the state of Colorado.

THYMIC PORTRAITURE

True insights into self don't come cheaply. A little like staring long and hard at the face of a watch and trying to think your way through the mechanism. Somehow it seems it must have to do with time itself, some inherent will of the watch to mark the passing moments.

But if you pry the face off the watch, you find that it isn't a bit of time stuck inside that makes the watch tick, it's a bit of trickery. Gears and levers, cogs and springs—trickery to simulate the passage of time. But unless you open the watch like an Oreo cookie, you'd never guess it.

It's a lot like that with humans, too, except we usually volunteer to disassemble ourselves.

———————

Out of biological necessity, each of us has a self. Though at times we might wish otherwise. Most of us imagine that self lives somewhere inside our nervous system, and especially our brain. But that is only part of the story, a part we will get to later.

One guidebook to the self comes from diseases like HIV/AIDS. The human immunodeficiency virus attacks human immune systems and in the process destroys our ability to deal with infections. Shortly, where one of us once stood, there rises a community of living things—bacterial, fungal, parasitic, viral, human, barely. The individual, the self, has evaporated, and in his or her place stands a murky mixture of living things that resembles a human being less each day.

Part of our selves has to reside apart from our brains and inside our immune systems. Right now, that's the part that is threatening to take Sandy's nervous system apart a piece at a time.

———————

How immune systems establish and protect selves is one of the most remarkable stories biologists have ever unraveled—even though much of the text remains untranslated.

Near the end of the third trimester of a human pregnancy, billions of lymphocytes (one group of white blood cells) leave the baby's newly formed bone marrow and head for the fetal thymus—an organ about the size of a large marble—that lies just above the fetus's heart. Inside the thymus, during the last three months of fetal development, a self is born.

It is then that most baby boys and girls learn to separate themselves from the rest of this hungry world. In the process, those lymphocytes become T (for thymus) lymphocytes.

Soon, it will be the job of these newly born T cells to defend the child from all the bacteria, viruses, parasites, and fungi that would like to farmstead the acreage he or she will learn one day to call "me." But before T cells can do that, a lot has to happen, and the way it happens places us all at risk for darker possibilities.

When the T cells from the bone marrow arrive inside the infant thymus, the thymus forces them to start dividing and to begin expressing a particular set of genes called T-cell receptor genes. The proteins that come from these genes are all we will ever have to sense the infectious world around us and react to it. Animals without T-cell receptors die young.

Inside thymuses, as the new T cells begin to divide, they rearrange portions of their DNA, actually chop out pieces of chromosomes and stick them randomly into other parts of those chromosomes.

The DNA passed to us by our parents tells a story that is over 3 billion years old. And that story has been edited and reedited over all those years into a very fine and fiercely accurate tale of what it means to be a living thing on this planet. Because of that, for any one of us to dare change even a single word of this story is dangerous.

But not inside the thymus. Inside the thymus, we treat our DNA as if it was a deck of cards. Inside the thymus, T cells pick up their decks of DNA and, in spite of billions of years of warnings, the T cells start to shuffle and reshuffle and reshuffle until they have generated something like 10^{15} different T-cell receptor proteins. Human minds can't handle numbers like one with fifteen zeroes after it. Still, it's clear that's a whole lot of T cell receptor proteins—essentially one for every infectious threat we might encounter in a lifetime and then some. And the "then some" is where things can get ugly.

With a deck of cards, after honest shuffling, there is no way to control the hands dealt from that deck. That's why some people lose everything they own playing cards. Same with T cells. After all the shuffling that goes on in the thymus, there is no way to control the sort of T cell receptors made. That's why some people lose their lives to their T cells.

"Random" means that along with protective T cells, every one of us makes T cells with a taste for self. The thymus has the job of trying to find those T cells and destroy them. And the thymus is pretty good at that. More than 90 percent of all the T cells that arise from cell division inside the thymus die before they get a shot at anything but the thymus.

For the good of the whole, the thymus bares its soul to the developing T cells so that every T cell that might react with any of the proteins in the thymus will show its true nature and can be culled and killed—it's a little like trying to separate the terrorists from the people's army.

Beyond that, the cells of the thymus have conjured a little magic that allows them to dress themselves up in costumes that look like heart cells, muscle cells, kidney cells, retina cells, liver cells, and on and on. In truth, because of its genetic legerdemain, inside of every human thymus there is a nearly complete portrait of self. A portrait painted by the thymus itself. And that portrait is spread before all the developing T cells and the T cells must choose. If they choose poorly, the thymus moves swiftly and executes those T cells before they can do any harm.

A nearly perfect solution to a nearly unimaginable challenge.

Nearly perfect.

Usually, nearly perfect is good enough. But not with thymuses.

In spite of the ruthless search for self-reactive T cells inside the thymus, a few always escape. Maybe they need to escape to keep us whole. Because of those thymic imperfections, each of us carries in his or her blood the means for self-destruction.

That is the price we pay for having selves. Our thymuses, at some considerable risk to us, create a certain sense of self. A powerful sense we could not survive without. A sense that is uniquely ours, a mosaic of DNA unlike any our parents ever held, a piece of the story that no one else has ever heard or told. Without that, that thymic sense of self, we would all have long ago succumbed to bacteria and viruses. An image of self held at once by trillions of T cells coursing through our veins and arteries and our tissues. 10^{15} possibilities, maybe more. If you counted seconds on a clock, it would take you roughly 500 million years to get to 10^{15}. That is the power of human immunity. But that near miracle comes with a price. And right now, in the United States alone, nearly 24 million people, like Sandy, are paying that price.

Autoimmune diseases offer one powerful view inside. But so do psychoses. MS offers a little bit of both.

PAT

I know many dead women. I know cemeteries full of dead women.

The day they buried Pat Brooks, I stayed home. It was wintertime, January maybe. But a nice enough day, as I recall. No obvious reason to stay away.

Pat and I had known one another for about fifteen years. I taught Pat immunology during her first year in vet school. And we got together regularly after that for as long as she was in school here in Fort Collins. When Pat graduated, she left Colorado and spent some time at the University of California at Davis. But two years later, I offered Pat a position as a graduate student in my laboratory. She accepted. For the next two years, we worked together as graduate student and mentor. And for all of that time, she was a good friend. Gina, my wife, and I often shared meals with her and the man she lived with, times in the hot tub, music, even a bike ride or two. It was, in part, because of me and the job I offered her that Pat returned to Colorado State University and Fort Collins where she married Tom Spurgeon.

Unknown to either of us, Pat had several things working against her, working against her in a way that would eventually kill her. Pat, too, was a woman;

she, too, grew up in Colorado; and she, like Sandy, was of northern European descent. On top of that, during the course of her life, she had—like all the rest of us—been infected by simple things that could unlock the dark secrets buried inside some of us.

Because of all that, Pat was in danger. But none of us knew that. So we carried on as though each moment was exactly like the one before, as though there were a million moments yet to come.

In spite of all we had shared, I stayed away from Pat's funeral because during the last couple of years, my relationship with Pat had deteriorated. It was, of course, something trivial, and it seemed like we would get around to resolving it one of these days. We never did. Anyway, I figured because of our falling-out that maybe she'd prefer I was somewhere else when her friends and family offered their last respects. I wanted to honor that perceived wish.

A few days before, Pat had purposely loaded a pistol and shot herself in the head. She was thirty-two, maybe thirty-three years old. Pretty, young. We all assumed that Pat had done that to herself because earlier the same day, her husband, Tom, died in a car wreck. A young man had run a stop sign at high speed and slammed into Tom's car, crushing the little Toyota and splitting Tom's aorta. He was dead before the last of the shattered glass hit the pavement. That seemed reason enough for Pat to choose to end her life. Nevertheless, that was not the reason Pat killed herself, or at least not the only reason.

At Pat's funeral, I was later told, her brother read from her journal. A journal that no one had known existed until then. In Pat's words, her brother revealed to everyone listening that Pat had been diagnosed with multiple sclerosis a few weeks before Tom's death, a few weeks before her own death. None of us knew about that. But through her brother, Pat told everyone at the funeral how devastated and confused she had been when she was told of the disease. And how she had kept all of that to herself. So the day Tom died, none of us knew that Pat's immune system was quietly eating away at her spine and brain. Not even Tom knew about it. When the last of Pat's friends left her that night, though, Pat remembered. She decided that that was enough for her, enough to know for one lifetime. And she put the final period into her story.

BRAIN WAVES

Of course, powerful and deadly as it is, it wasn't Pat's immune system that pulled the trigger the night of the automobile accident.

There is another self—one with eyes and ears and taste buds, one with fingers to touch our selves and one another. The simple self—the self of volvox and planaria—was quickly outsmarted. As things grew larger, selves that only understood "Don't eat me," "Don't speak to anyone else," and "Destroy anything inside of me that is not me," didn't have much to offer when megalodon (a prehistoric shark about seventy feet long) stopped in for lunch. Situations like that demanded relocation, rapid relocation, and maybe even the rudiments of thought. The stage (or the dinner plate) was set for muscles and nerves and brains—the other self. Not a simple scheme of secret handshakes like the first, but an elaborate system of wires and switches, triggers and fuses, levers and transmitters—nervous self. Just as immune systems had, nervous systems rescued our ancestors more than once.

In response, biology bestowed upon each of us a clot of mental and neurological phenomena that allows us to distinguish our bodily selves from all that surrounds us and react. Another, newer sense of self.

Nerves, neurons, and the signals they transmit are essential to some of the most complex manifestations of self and self-perception. Among humans, the greatest concentrations of neurons are in the brain and spinal cord. There, our thoughts and our actions collide—ideas become poetry or music, metal ingots or crescent wrenches, babies and vineyards—all through the magic of neurons.

Neurons are curious-looking cells. Each has a cell body, which houses the nucleus and the chromosomes, and arms that radiate off that cell body like a starburst. Each of the rays in that burst has a specific function. The longest and strongest is the axon, the others dendrites. Axons carry nerve impulses. Dendrites mostly receive the impulses delivered by axons. Neurons are the business cells of the central nervous system. All the messages to and from command central in the brain go in and out through neurons and axons.

Inside human brains, there are two very different sorts of neurons—those in which the axon is naked and those in which the axon is wrapped inside of layers of myelin. Myelin is a fatty substance that allows nerves to deliver signals more rapidly. In one sense myelin serves to insulate the axons, but not in the same way the plastic wrapping insulates wires in our homes. Myelin accelerates the nerve impulse by forcing the electrical signal to leap from point to point along the axon. As a result, a nerve impulse moves much more quickly down myelinated axons than they do down unmyelinated axons. Much of what makes us human and keeps us alive in a predatory world would be impossible without myelinated axons.

Inside the brain and spinal cord, these two types of neurons are responsible for the distinct appearance of the white and the gray matter. White matter consists almost entirely of myelinated axons, while the gray matter is made up mostly of unmyelinated dendrites and cell bodies of the neurons. The deep centers of the brain are mostly white matter, the outer cortex mostly gray matter.

The second piece of human selves is locked up somewhere in there, somewhere in among the neurons and their myelinated axons, somewhere inside our brains and along the twisted cables of our spinal cords.

Out of the mind's idea of self and the thymus's portrait of self come human selves and self-awareness. Human existence and human persistence. Science, poetry, religion, philosophy, song, hatred, war, love, bigotry, and brotherhood.

And so similar are the neurological and immunological processes that are involved, so close is the communication between lymphocytes and neurons, between thymus and lymph node and brain, that I think that these are not (as we have been told for so long) two unrelated systems, but a single functioning unit. A single system—neuroimmune—whose primary task is to speak the words of the only stories that are truly ours, the story of self and self-defense. A system to name us and to defend us from things as varied as viruses and vodka, parasites and patriarchs, bacteria and Buicks.

Inside of us are the words, and inside of the words are our stories. Those stories can save us or kill us.

PALIMPSEST

These are gifts from those who came before us, and they will be our gifts to those who follow. The gift of self and the power to defend it. But there is a curse as well, a curse laid upon the gift by the evil goddess of necessity. The curse states that in return for our precious selves and their defense, we must, at once, acquire the weapons for the destruction of those selves. And further, that any control we exert over these weapons is tenuous at best. Illusory at worst.

The magic of immune development is nearly, if not completely, beyond our imagining. Here cells mask themselves in the raiment of others, perform real magic, and carve, from the plain stones of sperm and egg, a human being and its antithesis—the worm of self-destruction.

At the same time, the complexity of neurological self speaks a tale so fanciful, so lurid, that we may never understand what separates the whole from the pieces, the firm from the infirm, the sane from the insane, or what takes human beings down any of the paths that we must follow. The gifts of imagination, courage, resolve, strength of will, and vision carry with them the inherent risks of depression, delusion, anxiety, insanity, and suicide—unavoidably.

Within our nervous systems, of necessity, there are the seeds of psychoses. Within our immune systems, of necessity, there are the roots of self-destruction. Self, healthy self, depends absolutely on both systems and insists as well on the flaws.

The Disease at the Seam

Multiple sclerosis is about the purity of our two selves and about their flaws. It is the mystery in the middle of the pool. Among all the autoimmune diseases, multiple sclerosis is, at once, one of the most remarkable and one of the most sinister—because this disease works in between the two selves. But as it does that, as it pries the two selves apart, MS opens a window on the human self.

Inside people with MS, the immune system believes the nervous system is the enemy. Because of that, the immune self lashes out at the nervous self. All the while, the nervous system is, of course, manning the controls of the immune system, orchestrating its own destruction. All the while the immune system is mucking with the dials and levers that regulate the nervous system.

The immune system strips a neuron of its myelin. The nervous system, in its turn, raises the voltage. The immune system reaches a little deeper. And the whole thing ratchets up a notch. Self versus self. A state of being quite unlike any other.

People who get multiple sclerosis have some things in common. Caucasians get MS more than any other ethnic group. Caucasian women are three to four times more likely than Caucasian men to develop MS. In general, women who live at or above the fortieth parallel north (a line that, in this country, runs from about the Empire State Building in New York through Columbia, Ohio; Springfield, Illinois; Fort Collins, Colorado; north of Salt Lake City, Utah, and Carson City, Nevada; to Point Delgada, California. People who grow up living on or above that line are twice as likely to develop MS as those living below it.

Similarly, people south of the fortieth parallel south—places like southern Australia and New Zealand—have a higher incidence of multiple sclerosis.

Interestingly and inexplicably, women who grow up in the state of Colorado are ten times more likely to develop the disease than women living in southern states. Girls and boys under the age of fifteen who move to Colorado become as likely as those born in the state to contract MS. But people who move to Colorado after age fifteen are no more likely to contract MS than their southern neighbors.

Most people with MS are descended from northern European peoples, especially Scandinavians, and many with MS have been infected by human herpesvirus 6 (the virus that causes roseola in children), or measles virus, or (especially) Epstein-Barr virus. Bacteria such as *Chlamydia pneumoniae* have also been found in much greater numbers in people with MS. The difficulty, of course, lies in distinguishing causes from effects in MS.

Occasionally there have even been outbreaks of MS, and groups of people have simultaneously developed the disease. Such outbreaks are called "clusters." Between 1943 and 1989, four separate clusters of multiple sclerosis outbreaks occurred in the Faroe Islands. This group of islands rises between Iceland and Scandinavia and was occupied by British troops during the Second World War. After that occupation, the number of cases of MS increased each year for twenty years. Other clusters have been reported among people living in the same neighborhood or people working in the same office. Sandy tells me the same seems true of the neighborhood where she grew up.

That's what we know, which isn't much. What we don't know is what causes MS, which is a lot.

Infection may be part of it. But it is not clear which infections are most important. Things as diverse as typhoid, mumps virus, and EBV might be important. But no "smoking gun" has been found. And how an infectious microorganism could cause a human autoimmune disease is murky at best.

The higher incidence of MS among people of northern European descent suggests that a person's genes play a role in determining whether or not that person gets MS. Interestingly, one of the candidate genes is the blueprint for a molecule that hands over suspected invaders to T cells. Maybe something gets mixed up in that process and a piece of self comes to look much like a piece of something a whole lot worse.

Self vs. self. In the ensuing battle, the myelin sheath is stripped from the neurons of the brain spinal cord (sometimes the optic nerve, sometimes olfactory nerves, sometime the nerves that enable us to walk and speak normally) and then the immune system attacks and kills the oligodendrocytes (the cells that make the myelin sheath) and some of the neurons.

When that happens, people change. Things no longer sound the same. Things no longer look the same. Gait may change, swallowing may be more difficult. Sometimes there is an incessant tingling in hands or feet, dizziness, loss of memory, or even a wish to have no children. Self, itself, changes.

Early on in multiple sclerosis, some of the time, the damage to the myelin sheath is repaired, and things return to normal for a time. But often, the disease flares again. As things progress, the immune systems often win out. The repair efforts cannot keep up with the wounded and the disease gets worse and worse.

Nerve transmission slows. Plaques—demyelinated patches of white matter— turn gray in the brain and spine. Missteps appear on the way from an office door to a classroom, speech curls in odd and unpredictable ways around familiar vowels and consonants, sometimes legs give out, sometimes eyes, ears, or noses just quit working.

Imagine what it must feel like to have the two most intimate parts of ourselves—parts most of us take completely for granted, the parts that make us who we are—warring with one another. Imagine self itself at risk. Imagine such a sudden and searing insight into the elemental fire of our individuality. An image uncluttered by all the confabulations of life, unobstructed by the nonessential. An immaculate vision into our heart of hearts. How that must feel.

SANDY

"Unfair," Sandy says again, unhesitatingly. But "for a purpose, a reason," she says. Nothing happens without a purpose. Sandy's certain of that and willing to wait for it, even if while waiting she loses most of everything she ever thought of as her self.

Sandy's courage frightens me. She is human, like us all. But she has found another story—one with fewer words and fewer reasons. And inside that tale, in there with the anger and the hope, the courage and the fear, there are words of encouragement for others like me.

The only road left open to the two us is the dark road that leads inside. Sandy, with her slender hands and her gentle smile, is leading me a little farther down that road just now. All that I have learned in my books and my laboratories is useless now. This is not an intellectual struggle. This is a fight for survival.

"Has it changed the way you see yourself?"

"I've thought a lot about that. And yes, it has changed things. I feel like a piece of me is missing. I don't feel like a whole person anymore. I'm not like everyone else anymore," Sandy says with her flinty honesty.

We learn quickly what it means to be diseased. How it sets us apart from others, others with their self-confidence, their health, their corrosive innocence. We learn that nothing is the same anymore, and we notice that people look at us differently, if they choose to look at us at all.

But disease does open windows, windows that frame sights no one else has ever seen, delivers us onto rivers that carry us to places our health denied us. Pictures visible only to weakened eyes, places and people touchable only with tingling or broken fingers, thoughts plausible only to demyelinated or crushed neurons. A passage, paid for with human flesh.

Sandy is showing me that way.

I'm afraid.

"Does it hurt?" I ask.

"Big time," Sandy says. "Big time."

El Santuario de Chimayo

3

Self in the Soil

Please
This earth is blessed
Do not play in it

—*Sign on the wall of El Santuario
de Chimayo, New Mexico*

I'm not one for religions or religious experiences. But there is something here I cannot account for—something very old and very unusual. The carvings and paintings are part of it. They were surely done by human hands, but according to public documents no one remembers whose hands those were. The work is striking, especially in the apse behind the altar. There, the colors of the surrounding hills have been transferred onto nearly luminous wooden reredos full of Catholic symbolism. Above the altar, a blackened wooden Christ hangs crucified on a green cross. And over Christ's head, the roof is held in place by carved wooden beams, big around as human bodies and darkened from more than two centuries of incense and candle smoke.

DOI: 10.5876/9781607322337.c03

The air reeks from thousands of benedictions and baptisms. Ten or twelve rows of high-backed, wooden benches and kneelers fill the nave. The bottoms of those benches glisten from decades of frayed trousers on fidgeting children and ripple with a low orange glow from dozens of votive candles burning purposefully at the back of the church.

This is El Santuario de Chimayo, an old adobe-brick and stucco structure in the hills of northern New Mexico. This chapel was built in 1816, but a sanctuary has been at this site for much longer. The locals offer many legends about its origins, fanciful tales of miraculous crucifixes and Santo Niños. The truth, if there is one, is buried beneath the murk of time. But as mystical and beautiful as the sanctuary is and as striking as the crucifix (El Señor de Esquipulas) above the altar is, nearly none of the supplicants in these pews today have come to see the sanctuary or the crucifix. Instead, they have come from all over the world to this place in New Mexico to eat the dirt that lies beneath the adobe floor.

According to legend, that dirt is sacred, consecrated by Christ himself. Crutches cast off by the newly healed fill the anteroom, and on some days, the line of pilgrims stretches for blocks.

To see what the faithful have come for, you must walk to the front of the church and just beyond the Communion rail, there, to the left, lies a low-ceilinged entrance to a small room. There, a hole (the *posito*), about eighteen inches across, pierces the floor of the church. Beside it, someone has left a plastic spoon. Inside the posito lies the deep-red dirt of Chimayo. That dirt tastes of grit and time and something else—maybe cinnamon or cumin seeds. Maybe salvation.

Shops that surround the santuario sell plastic bags and wooden boxes, bronzed images of the church with hinged tops, and glass vials stoppered by tan corks—all created to contain a few plastic teaspoonfuls of that dirt. Few of those who come to this church leave without it.

Religious dirt eating, so far as I know, is practiced at only one other place in the New World—a Catholic shrine in Esquipulas, Guatemala. But pilgrims to Chimayo and Esquipulas are not the only humans who eat dirt. Nor are religious reasons the only reasons to imagine that dirt may have special powers.

DIRT FOR DINNER

> A few drops of Tabasco, and you've got a meal*
>
> —*Anonymous e-mail*

> *Dis-moi ce que tu manges, je te dirai ce que tu es.*
> [Tell me what you eat and I will tell you what you are.]
>
> —*Anthelme Brillat-Savarin, Physiologie du gout;*
> *ou, Meditations de gastronomie transcendante, 1826*

Much of who we will become we acquire from the world around us, includ-ing—in fact, especially including—dirt. Other than water, what little stuff we humans have inside us is largely dirt. Admittedly, this dirt is sometimes highly processed before we receive it: in the form of cows and sheep and carrots and squash and bison and sorghum.

But not everyone wishes his or her dirt to be so far removed from the stuff of mud pies and mucilage. On every continent (except, possibly, Antarctica), some of us intentionally eat dirt, and we are joined in this practice by myriad rats, mice, mule deer, birds, elephants, African buffalo, cattle, tapirs, pacas, and several species of primates.

The majority of scientists view most animal geophagy (dirt eating) as nor-mal. Probably because we consider all animal behavior to be "natural," without thought or perversity. However, most of these same scientists consider human geophagy abnormal.

ABNORMAL (?)

> I am a (going on 51 in October) woman that has been eating dirt
> since I was 3 years [old]. I always thought I was alone in this eating dirt
> thing, but I see I'm not alone. I guess I've been eating dirt for the great taste

---

* Asterisks identify quotations taken from e-mailed comments I received after publishing a first version of this essay in *Emerging Infectious Diseases*, a publication of the Centers for Disease Control and Prevention in Atlanta, Georgia.

and the way it melts in my mouth. Dare I say that it's even better than sex[†]
(and believe me I have a great sex life). I am picky. I grind it down. Then, I
sift it to where it's very powdered down. I must eat about 15–16 lbs a week.

Sometimes more.*

—*Anonymous e-mail*

In the United States, many of us believe that humans should eat only food.
Because of that, we consider the consumption of nonfood items pathological,
even though we know that what people define as "food" varies dramatically
both ethnically and regionally. We call eating nonfood stuff "pica." And since
soil does not appear on the Federal Drug Administration's food pyramid, we
define soil pica as a disease.

But there are rules. All dirt eating is not soil pica. It all depends on how much
dirt a person eats and whether he or she eats that dirt by choice or through an
act of God.

The authors of one report I read described soil pica in a developmentally dis-
abled person who regularly consumed more than fifty grams (that's about two
ounces, or two teaspoons) of soil per day. Because of some (apparently unwrit-
ten) standards, most in the business of evaluating dirt-eating folks seem to agree
that this level of geophagy is pathological, although I am not sure why.

Apparently, neither was anyone else.

So in June of 2000, the people at the US. Agency for Toxic Substances and
Disease Registry appointed a committee of men and women to review soil pica.
The committee settled on pathological levels as consumption of more than 500
mg (about 1/10 of teaspoon) of soil per day. But the committee members themselves conceded that the amount selected was arbitrary.

In truth, especially when the wind is blowing, most of us consume more than
1/10 of a teaspoon of dirt every day. That would seem to make the declaration of
the US. Agency for Toxic Substances and Disease Registry worse than arbitrary,
and nearly useless.

[†] Since I have never met this person, I assure you that she did not mean that eating dirt
     was simply better than sex with me.

The only human soil consumption considered by some to be even slightly normal is dirt eating by pregnant women (especially in sub-Saharan Africa), migrants from sub-Saharan cultures to other parts of the world (notably the United States), and children worldwide.

Normal, abnormal, and unavoidable, the soil we eat is a large part of who we are and who we will be. The soil eaten by children may even be an essential part of becoming human.

TRADITION AND CULTURE

> When I was a teacher in Botswana, my female students used to sneak dirt into my class to eat from under the desk. It seemed a gendered behavior—boys never did it. It was normative there for pregnant women to eat dirt (especially dirt from termite mounds) and my interpretation was that one of the reasons the girls might have been eating a dirt was because it was a signifier of being/becoming an adult woman.*
>
> —*Anonymous e-mail*

For centuries, indigenous peoples have routinely used clays (decomposed rock, silica and aluminum or magnesium salts and absorbed organic materials) in food preparation. The clays were used to remove toxins (for example, in Native American acorn breads); as condiments or spices (in the Philippines, New Guinea, Costa Rica, Guatemala, the Amazon and Orinoco basins of South America); and as food during famine. Most recently, dirt became a commercial commodity in Haiti in the form of soil cookies or biscuits sold by roadside vendors to desperately hungry men and women. Clays are also often used in medications (for example, kaolin clays in Kaopectate and bentonite clays in antifungals). But the most common occasion for eating dirt in many societies (the only occasion in some societies) is pregnancy. In fact, in several areas of this world dirt eating is diagnostic of pregnancy. When sperm and egg collide, the world changes. That is obvious. But why pregnant women eat dirt is not.

Eating clay, for several reasons, could be useful for pregnant women. These clay soils contain a lot of kaolin and that could ease gastrointestinal upset common in the first trimester. Also, kaolin clays absorb toxins, some of which might come from plants in these women's diets, though that wouldn't explain why they eat dirt only when pregnant. And finally, these clays contain calcium, an essential element for building baby bones.

But not just any dirt will do. In some cultures, well-established trade routes and clay traders make rural clays available on city street corners, because these urbanites prefer the same soils their ancestors ate. And clays from rural termite mounds are especially popular among many native peoples. Termite mound soil is rich in calcium.

It isn't known when and why these particular sites and sources of dirt became popular. Nor is it clear just when people started eating dirt, but some anthropologists say it began before *Homo sapiens* appeared on this Earth. That means that we and our ancestors have likely been eating dirt for more than a million years.

Also, the times for eating dirt vary among different peoples. In some places, women may eat dirt only during the first trimester. In other locations, dirt is eaten in just the second or third trimester. In still other countries, women eat dirt throughout pregnancy. Most of the women who eat dirt eat the soil throughout the day as a supplement rather than a meal.

The most popular of the available dirts are subsurface clays. These soils disappear into pregnant women at the rate thirty to fifty grams a day (about a quarter cup), although some eat much more. If this dirt was a major or sole source of calcium, you might expect that tribes with dairy herds would eat less soil. That doesn't seem to be the case. Also, because of my interest in dirt, I have come in contact with a dozen or more people in the United States who eat dirt. The reasons they have offered for their hunger are various, but one theme appears in all their stories—the dirt just looks and tastes good to them. Nutritional need, or even a sense of need, has never been mentioned to me as a reason real or imagined for eating dirt in the United States.

So if it isn't nutrients dirt eaters are after, what is it?

The short answer is nobody knows.

Here's one possibility. Soil contains considerable amounts of organic material, including many live microorganisms—especially bacteria and fungi. The human gut is the largest area of direct contact between a person and the world. To function properly, that gut needs to house billions upon billions of bacteria.

Also, major portions of human immune systems do their own special voodoo inside human intestines. In fact, these spots in our guts even help to generate some of the white blood cells we couldn't live without—which makes human gastrointestinal systems mysterious swamps where nearly magical things regularly occur.

Among those nearly magical things is a special type of antibody called IgA. Antibodies protect us from all sorts of other creatures, many of which would turn us into food if they could. Because babies can't make antibodies as fast as adults, infections regularly take a very heavy toll among human infants. As recently as 1900, more than a third of the babies born in the United States never saw their fifth birthdays. Mean microscopic creatures with a taste for human flesh—and all that stands between them and us are our immune systems, especially our gut-associated immune systems.

Things we eat often immunize as cleanly and as quickly as syringes and needles. And that immunity is regularly passed on to fetuses and babies across the placenta and through breast milk. So maybe eating dirt immunizes moms and then they pass that immunity to their babies, before and after birth. In support of this, it appears monkeys that regularly eat dirt have lower parasite loads and healthier babies.

In some human cultures, people bake the clay before eating it. Curiously, that is the very same thing we do with many vaccines—the so-called heat-killed vaccines. The baking is done to eliminate the microorganisms' power to reproduce but not their capacity to induce immunity.

Furthermore, for decades we have used aluminum salts—like those found in clays—to enhance the potency of human vaccines, because they stimulate inflammation, and that makes for better immunity.

In pregnant women, this sort of immunity could offer protection against precisely those infectious threats the baby will face at birth. Eating dirt, then, rather than being abnormal, may be an evolutionary adaptation acquired over millennia of productive and not-so-productive interactions with bacteria—an adaptation that enhances fetal immunity and increases calcium, eliminates gastric upset, detoxifies some plant and animal toxins, and perhaps boosts mothers' immunity. And all of this at times when the hormones of pregnancy, factors produced by the fetus, changes in the serum and cell-surface proteins, and who knows what else suppress the mother's natural immunologic desire to destroy her fetus—a miracle, nearly.

Born Hungry (?)

> From Uganda: my daughter started this habit at the age of two
> years and I was surprised by this and I started giving her some deworming
> tablets. So what side effects does this have?*
>
> —*Anonymous e-mail*

My children ate dirt with surprising gusto: garden soil, road soil, leaf-mush soil, sunlit soil, moonlit soil, sod soil, bug-body soil—even gutter soil. As usual with my children, before I could talk them out of this behavior, they gave it up on their own—their behavior depending more on personal likes and dislikes than on my paternal concerns. I was pleased when they quit. Later I was reassured to discover from other parents that their children were just as taken with dirt as mine were, some even more so. I felt less like the parent of a couple of dirt-eating, psychosis-ridden, nutritionally deprived children, even if my children, like me, were never quite "normal."

Eating dirt is nearly universal among children under two years of age. When I asked my two-year-old daughter why she ate dirt, she just stared at me, her eyes wide open, a thick moustache of loam limning her lips. She must have decided either that I had asked something unfathomably abstract or simply that my question was too nonsensical to warrant an answer. So she just resumed eating dirt.

Soils consumed by children may differ from those consumed by adults. Generally, children consume topsoils and not the deep clays adults prefer. And children are considerably less selective about where they get their dirt.

Children may eat soil for some of the same reasons pregnant women and some animals do. Because of their rapid growth, children have special nutritional needs and surface soils may help them meet some of these needs; detoxification of plant or animal toxins might be accelerated by geophagy—particularly in some parts of the world; or soil components, especially clays, may relieve gastric distress. But topsoils are probably not as effective at gastric soothing as are deep clays.

Among children, too, it seems possible that eating dirt could have immunologic consequences. Some maternal antibodies cross the placenta to the fetus before birth. Others appear in breast milk shortly before birth and for a year or more afterward. These antibodies provide much-needed protection in the newborn and last about eighteen months to two years.

Children often begin eating dirt a year or two after birth. As maternal immunity wanes, eating dirt might "vaccinate" children who are losing their mothers' immunity. That dirt may stimulate these two-year-olds to start making their own antibodies, especially IgA. Eating dirt might also help populate intestinal flora—the bacteria in our intestines that are essential for normal gastrointestinal and immune development.

The Environmental Protection Agency estimates that children in the United States consume, on average, 200–800 mg of dirt per day. Some children regularly consume more than their allotment. Still, while plenty for immunization, that doesn't seem like a lot of dirt. Nevertheless, we parents have tried for years to put a stop to it. I don't know of an instance in which anybody has succeeded in keeping children altogether away from dirt.

But animals have been successfully raised in absolutely sterile environments. Rabbits, mice, guinea pigs, and rats have been raised under such conditions. In each case, these animals' immune systems developed abnormally. Lymph nodes (little peanut-shaped organs where immune responses develop) and gut-associated immune sites were malformed and incapable of normal immunity. Reexposure to infection later in life did not repair this damage.

Apparently, there is a window when infection drives the immune system and the gastrointestinal system toward their proper ends. If that window closes before massive infections populate the gut and immune systems, mice, rats, rabbits, and guinea pigs are at the mercy of the microbial world.

It is likely that we mammals are equally dependent on early and massive infection. Children with many older brothers and sisters are less likely to have asthma, hay fever, or eczema. West African children who have had measles are half as likely to have allergies as children who never had measles. Italian students who recovered from infection with hepatitis A had fewer and less severe allergies than fellow students who were never infected. Children with type 1 diabetes (an autoimmune disease) are less likely than healthy children of the same age to have had infections before their fifth birthdays. Children raised in rural areas, especially on farms, have fewer allergies and autoimmune diseases than children raised in cities.

Children exposed to a little more of the infectious side of this world seem to fare better as adults, to less frequently recognize certain innocuous organisms as a threat, and to more readily discriminate between self and not-self and separate the fatal from the innocuous.

RISKS OF EATING DIRT

> I am very concerned about my 21-month old daughter's habit of eating sand, particularly wet sand. A few months ago she was pretty indiscriminate, eating dirt from the garden or sand from the playground. I thought that she had moved out of that stage, until her caregivers at day care told me that she had been eating sand again. Her diapers are terrible to change, poop loaded with sand that has passed through her system. It hurts her to wipe her (like wiping her with sandpaper!) and she is getting a rash.*
>
> —*Anonymous e-mail*

How dangerous is eating dirt? My mother was pretty certain about this—damn dangerous. These days, it's hard to find plain old earth. Most of the stuff we walk on has soaked up many years of human wear and tear. Sometimes that includes PCBs, mercury, lead, and who knows what other by-products of human progress. Because of that, you need to be selective about what soils you sprinkle on your morning cereals or even, depending on where you live, quit eating dirt altogether—although the wind makes that pretty near impossible.

The inherent biologic danger of soil is difficult to assess. Soil unaffected by the pressures of overpopulation, industry, and agriculture may be vastly different from the soil most of us encounter routinely.

Using DNA-hybridization analyses (a way to measure the amount and type of DNA), scientists found about 4,600 species of microorganisms per gram (1/28 of an ounce) of natural soil—mostly bacteria and fungi. That's a lot of variety. Other investigators, using more sophisticated techniques, were able to measure both the variety and the mass of living things in the dirt. These investigators found even more species and 0.7–7 kilograms (about 1.5 to 15 pounds) of biomass (living and dead micro- and macroorganisms) per cubic meter (about 1.5 tons) of soil.

Since none of us is likely to eat a ton and a half of soil, it might be more useful to think of this as 3.5×10^{-4} grams per teaspoon. About 3/10,000 of a gram of recently or currently living organisms in a teaspoon. That seems pretty trivial. But since the average weight of a bacterium is about 7×10^{-16} grams, that teaspoon of soil could hold about 400 billion (400,000,000,000) bacteria, which by anyone's standards is lot.

Our soil literally teems with living creatures. In spite of that, there is little evidence to suggest that eating or breathing dirt regularly causes disease in people.

The few infectious diseases directly tied to eating dirt appear in children, who as mentioned mostly eat topsoils. Certainly, the outermost crust of our Earth is the part most regularly contaminated by what we and our animals do. Because of that, eating these soils may be riskier, especially in certain areas.

One recent report describes the infection of two children (at separate sites) with raccoon roundworm (*Baylisascaris procyonis*). This parasite does terrible damage to nerves. Both children suffered severe neurological damage, and one died. The children ate the roundworm when they ate soil contaminated with raccoon poop. So eating dirt can have serious consequences, but it seems that consequences this severe are rare.

The most common parasitic infection among dirt eaters in the United States is toxocariasis—caused by a roundworm. Many domestic cats and dogs carry this roundworm. A lot of these animals show no or only minor symptoms, but they pass millions of eggs every time they defecate. And the eggs can persist in the soil long after the poop is gone. If a child eats that soil, he or she may develop a serious disease that can cause blindness. In the United States, *Toxocara* infects about 10,000 people every year. As a result, this worm currently infects about 14 percent of the U.S. population. Most of those infections occur before age four. Outside of the developed world, toxocariasis is an even bigger problem. So soil contaminated with dog or cat feces provides a major route for movement of parasites from pets into humans. Most of these infections result in few or no symptoms. But some have severe consequences. Every year, nearly 700 people lose their sight in one or both eyes because of a *Toxocara* infection.

Neither raccoon roundworms nor *Toxocara* are a regular part of soil itself. But clearly knowing what contaminates the soil that you or your children eat is critical.

CHIMAYO

> I found this very interesting. This opened my eyes up
> to these different ideas. But not good enough for me
> to run out and eat a handful of dirt.*
>
> —*Anonymous e-mail*

Here beneath the old wood and marvelous crucifix, I watch the faithful leave. I marvel with them at the miracle beneath this adobe floor, the same miracle buried beneath most every place human feet have trod.

Manufacturing a human being takes a bit of dirt. In truth, we are creatures of starlight and dirt. And though we most often take it wholly for granted, the dirt we walk upon, sit upon, and sometimes toss our trash into is sacred. Please play in it.

Holy dirt is offered free to those in need. However, if you would like to make a donation, please send a check or money order payable to "Santuario de Chimayo" and mail to:

P.O. Box 235
Chimayo, NM 87522
http://www.elsantuariodechimayo.us/holydirt.html

4

Gathering Our Selves

Though none of us begins that way, within the first few
months every normal healthy human being is (by cell number)
more than 90 percent bacteria. How can that be?

LUDWIG VAN BEETHOVEN
February 1824, Vienna

It is snowing, perhaps, and cold, surely. The streets are nearly empty.
The sparkling flakes that fill the stone crevices are radiant with candle
and gaslight. In a small villa near the city's center, at Beatrixgasse-
Ungargasse 5, a white-haired man in his fifties has just penned the final
notes of what will someday be called the greatest piece of music ever
composed. Beyond his windows, all of this goes unnoticed by the glittering crystals and the few men and women still moving through the
snowy streets. But the moment will be remembered forever by millions
of other men and women still unborn. The light dims, and the man
turns toward his bed.

DOI: 10.5876/9781607322337.c04

An integration of self-reflection after Rodin

Ludwig van Beethoven has just completed his final symphony, and it is a masterpiece. Such a masterpiece that, even today, every one of us immediately recognizes the grand finale of the Ninth Symphony. Because of the splendor of that music, most of us would agree that Beethoven was a genius. But where did that apparent genius come from? In truth, Ludwig had help—especially that night.

HENRY PERRY

June 1913, Paris

To the best of my recollection, I have known only one truly crazy person. Known well, that is. I, like most of us, have had more than one opportunity to observe the actions of the truly crazy from a distance—Dick Cheney, for example. But Dick and I were never on speaking terms, never even friends on

Facebook. My mother, of course, eventually achieved craziness; perhaps my father did, too. But they don't really count, because at one time or another all of us consider our parents to be insane.

Henry Perry was one of my mother's older brothers. He was born in Kansas in the 1890s—third or fourth in a family that would someday total thirteen. Unlike some, Henry wasn't born crazy. It was a thing he acquired along the way. Nor did Henry achieve craziness on his own. Like Beethoven, Henry had help. Henry came by that help one night in Paris.

About 100 years after Beethoven conspired with others to write his Ninth Symphony, Henry Perry sits smoking a cigarette at a small table in a rundown bistro in Paris. Across the table from him is a surprisingly beautiful woman. Henry has seen nothing but trench foot and mustard gas for the last months and is completely taken with the scent and the looks of the woman across from him. The woman, of course, is a whore. That makes no difference to Henry. Her breasts are magnificent, her breath cinnamon and cloves. *I've earned this,* he thinks. A simple wish for a single night of pleasure.

Henry is young, blond haired, and tall. The weeks at war have thinned him, and now he is shaped more like a man than a boy. His uniform, by some monumental accident, fits him perfectly. Tonight, Henry's long arms and legs seem just right to him and his escort. And Henry is sane.

Perhaps the waiter offers more wine. Perhaps they accept. The evening lengthens into darkness. Henry's desire gathers heat. Surprisingly, especially to her, the woman's interest is piqued as well. They move closer to one another. But that's what people do here, sit in dim corners beneath dark wood and eat from one another's spoons, like lovers.

Neither of them expects any of this to last beyond that night. That is what they have planned. That is what they have agreed upon. But there are others here this night, making plans of their own.

The Wrath of God

By the time I met Henry, he was completely insane. He couldn't speak in complete sentences. He couldn't walk the fifty feet across our backyard without jerks and staggers. Sometimes he drooled.

Henry was living, then, at the VA hospital in Salt Lake City, Utah. Every Sunday afternoon, my father would gather him up in our family's Ford station wagon and bring him out to our house in Bountiful for dinner. Henry seemed to like that.

His eyes were still piercingly blue, his hair still blond, though mottled with gray, and he still had the whippet-thin frame of a soldier. But the rest was no longer Henry.

In spite of his peculiarities, or maybe because of them, I enjoyed my uncle. He cursed and spat and wore soiled clothes, all of which I admired. But clearly something was different about Henry. Henry wasn't working alone either.

The voices that rattled in Beethoven's deaf ears that night in Vienna were not human voices. Mercury, from the vapor baths for his lifelong syphilis, silvered nearly every tissue in his body. Lead from his water or his knives and spoons was inside his neurons redirecting traffic. But, most important, the syphilis bacterium itself was in his brain and his spine playing its own music. Continual, bone-splintering pain worked at every nerve ending as Beethoven sat down that evening. Before the Ninth Symphony was finished, each of these others had its say. The Ninth was a cooperative effort.

A part of that same music was playing inside of Henry's head the day I met him in Bountiful, Utah. A tune with a ragged edge. In its own way, Henry's craziness was nearly as perfect as Beethoven's Ninth Symphony. I don't know if my uncle had ever heard of Beethoven, but a thing that had once been Ludwig's was now Henry's.

THE INFECTIONS THAT MAKE US HUMAN

Inside each of us a symphony is playing itself out. Syphilis is only one player in one of those symphonies. But because of the shrill note this bacterium plays, it gets a lot of attention. And because of that, the word *infection* carries its baggage like a handful of maggots. It shouldn't. Infections make us who we are and connect us to one another in the most intimate of ways.

Besides the bacteria that haunt us, there are others—gathered from seat cushions and kisses, from forks and knives and cheese rinds, and from mothers' wombs and breasts. Beyond the howling of *T. pallidum,* the syphilitic spirochete, others play a more subtle tune, and these others we couldn't live without. We have sought them out in the some of the lowest places imaginable, but they sustain us.

Bacteria rule the world of living things. All by themselves, bacteria account for well over 99 percent of all organisms—just as they have for all of time. Bacteria numbers are truly staggering—10^{29} on, in, and above the Earth; 10^{14} per person. If we round up the number of human beings on this planet to 10^{10}, bacteria outnumber us by a factor of 10^{19}—a lot. Inside our own skin, we carry around about 10^{13} human cells. Even within the space we call us, bacteria outnumber our cells by a factor of 10. Each of us, by cell number, is roughly 90 percent bacteria.

Clearly, we don't begin life that way. No bacteria shared our mothers' wombs. But by the time we are just a few months old we are lousy with bacteria. How does this happen? How does each one of us come to be so massively infested with these microscopic vermin?

PROMISCUITY

More than once I asked my mother to explain Henry's peculiarities, but it wasn't until long after his death that she told me the truth. Henry had syphilis. For my mother, that was a like slap in the face. Syphilis was a disease of the poor, the deviant, the unwashed. It was a sickness that fell upon the godless as punishment for their sins.

Syphilis, of course, isn't punishment for anything. It is simply a bacterium— *Treponema pallidum*—trying to make a living. *T. pallidum* moves from person to person during the most intimate of human acts. Wounds, torn tissues, cracked skin are all open doors for syphilis.

One night in Paris was all it took for Henry. Well, not *all.* It also took a partner and more. That night, as Henry ogled the young woman, someone invisible to everyone else was in the bistro. A creature who knew full well that this was not to be a one-night stand. The next morning, as Adrienne stayed behind in her small flat overlooking the rue Michelet, the spirochete, almost magically, both stayed with Adrienne and left with Henry.

Fleming's penicillin wouldn't come along for decades, so over the years, *T. pallidum* had its way with Henry. First, there was a minor wound, a chancre, not at all painful. And then it went away. Henry was relieved. Later, a rash spread across Henry's palms and the soles of his feet. He assumed it was left over from the trenches of France.

Headaches followed. Then the spirochete took Henry's joints—his knuckles, his knees—then it took his eyes, his spine, and his mind. When there was nothing more to take, *T. pallidum* took my uncle Henry's life, just as it had Beethoven's. It is even possible that Henry's syphilis could be traced in a direct line to Beethoven's syphilis, as clearly as any genetic link. After all, syphilis infects only human beings. Because of that, every syphilitic person is tied irrevocably to a long line of other syphilitics—reaching centuries beyond Beethoven and Vienna. To some that may sound horrible. But in fact the same can be said for millions of other infections thriving inside every living person. Our infections link us to one another as tightly as our humanity. Undoubtedly, Henry's lover paid just as dearly for her indiscretion, as did some of her previous lovers, and so on back to Beethoven and beyond.

Usually, we only really notice that—how our infections link us to one another—when the infections maim and kill us. But it happens all of the time. A touch, a breath, a spoon, a handrail, a kiss, and bits of us move from one to another. A shared bottle of spring water or a meal and we are no longer who we just were. Abruptly, I am you and you are me and we are all together.

And it all begins before we are even fully born.

BETRAYED AT BIRTH

At the dark moment of fertilization—when, with a final flash of its pearly tail, a sperm penetrates an egg—we are more human than we will ever be again. Sperm, egg, zygote, blastula, embryo—human. And for the next nine months, nestled in the sterile seas of our mothers' amniotic fluid, we remain mostly human.

But as birth nears, things change. In anticipation of our arrival, our soon-to-be mothers begin preparing a special nursery. Inside these women's birth canals, bacteria sprout like weeds. *Lactobacilli*, the same bacteria found in yogurt and buttermilk, divide and spread throughout the passage that we must traverse as we enter this world. About 1/100 the thickness of a human hair and as long as

1/25 the thickness of this page, these bacteria rise like fingers to prod us into reality.

On our way into this world these bacteria immediately infect us. Before our mothers have shared so much as a single caress, they have inoculated every one of us, infected us with billions upon billions of squirming, wriggling, living bacteria.

And it appears they must. If lactobacilli don't flourish inside a mother's vagina, premature delivery happens more often than in infected mothers. And infants born prematurely or by cesarean section face many challenges that full-term and fully infected infants don't. So everything conspires to ensure our immediate infestation.

After birth, things get dramatically dirtier. Our first breaths, the arms of the doctor or midwife or forest floor—all are teeming with microbes. Even in the relative sterility of a hospital delivery room, we roll in the powdered sugar of this world like a warm doughnut fresh from the oil. And as we do, we are quickly covered with layer upon layer of bacteria and fungi and viruses and even a few parasites. Life as a separate entity is over. From this moment on, no one of us ever walks alone.

As a mother nurses her child, she lays the groundwork for further infection. The milk she feeds her child is laced with proteins. Some of these proteins are fertilizers for more infectious microbes, especially *Bifidobacteria*.

The proteins in mothers' milk enrich the soil of the newborns' intestines. In that soil, *Bifidobacteria* push aside a few of the Lactobacilli and attach to the baby's intestines. Together these two bacteria weave a blanket inside the child, a protective blanket.

Children with too few bacteria often develop a disease called oral thrush—a yeast infection of the mouth. Without protective bacteria, a child's mouth sprouts thick white crusts of yeast across cheeks and gums and lips. As it spreads, the yeast digests the human tissues beneath and causes painful destructive ulcers. Without bacteria, life is harder.

The process of our infection appears completely chaotic, but it isn't.

Lactobacilli and *Bifidobacteria* come first. Then, as the mother withdraws her milk, the *Bifidobacteria* give way to other strains of bacteria, and much later, as the hormones of puberty flood the body, still other bacteria arrive and thrive.

Layer upon layer is laid down. Ordered and structured by time and chemis-try, we absorb our surroundings. Literally, we become what we eat or drink or touch or breathe. We are what we wear. We become those who caress us and those whom we caress in return.

From the world that surrounds us, we gather ourselves. Once pure and sterile, filled only with human cells, we transform into a microbiological metropolis cov-ered with living things whose names read like a Linnaean litany—*Staphylococcus aureus* and *epidermidis*; *Streptococcus mitis, mutans, viridans, pyogenes, and pneunominiae*; *Trichomonas tenax*; *Candida albicans*; *Hemophilus influenzae.*

Our skin sprouts a cornucopia of microorganisms, including nearly 200 dif-ferent strains of bacteria and several species of fungi—within human skin as well as upon it. And bacteria are not distributed uniformly across human skin. Some bacteria prefer the navel, others the forearm or underarm, and so on. And if you move bacteria from the forearm to the belly button, they don't last. In a short while, the original geographic pastiche reestablishes itself. Human skin is a mosaic of bacteria, each piece with a deep sense of place and purpose.

Human eyes gloss over with three or four strains of bacteria. Noses, throats, the upper reaches of our respiratory systems blossom with more than six differ-ent types of bacteria. Mouths cultivate a half dozen species of bacteria and fungi. Lower urethras fill with more than ten different bacteria, a few fungi, and a parasite or two. But by far, the greatest numbers of bacteria settle and prosper in our intestines. In places, the bacterial coat in our large intestine is an inch thick. And our feces, by dry weight, are 50 to 60 percent pure bacteria.

While no one notices, and with little or no effort on our part, we become a menagerie, a walking ecosystem, a universe apart—10 percent human, 90 per-cent other (maybe).

And it isn't just the number of microbes that is staggering. When scientists sequenced the human genome, they found only about 30,000 different human genes strung out along forty-six human chromosomes. But the bacterial genes we gather far outnumber the genes inside every human cell given to us by our mothers' eggs and our fathers' sperm. Amazingly, nearly 99 percent of the genes inside human beings aren't human. Perhaps even more amazing, inside each of us the mixtures of bacteria are as individual as human fingerprints.

INFECTION AND INDIVIDUALITY

Surprisingly, the community of bacteria within or upon one man or one woman is not simply an accidental consequence of birth and geography.

Where we are born is unquestionably important. Infants born in developing countries acquire bacteria that differ considerably from those of infants born in developed parts of the world. Children born in different hospitals may have very different strains of bacteria in their intestines. And breast-fed babies' intestines contain mostly *Bifidobacteria,* while formula-fed babies' intestines have more potentially dangerous bacteria, such as coliforms, enterococci, and *Bacteroides.* Where we are born and what we eat do make a difference. But our surroundings and our food are not the only factors.

No two of us are more alike than identical (monozygotic) twins. These children have the same chromosomes and the same human genes. Usually, identical twins also live in the same home, eat the same food, breathe the same air, and drink the same water. But there is one way they differ dramatically—each twin has his or her own, individually tailored set of bacteria.

Beyond monozygotic twins, even within a single geographic area, the species of bacteria found inside people vary dramatically from person to person. This is equally true of married people and families living in the same environment. You and your spouse, partner, brother, or sister house significantly different collections of bacteria.

She might have a little more *Staphylococcus aureus* in her nose or vagina than you have. *Candida albicans* might find him a little more attractive than you. *Helicobacter pylori* (part of the cause of peptic ulcers) makes a living for itself in some stomachs and small intestines but not others. *Citrobacter* (which can cause diarrhea and perhaps meningitis) is comfortable with some of us but not others. Our collections of bacteria are more individual than our fingerprints.

Many Japanese people even have special bacteria that millennia ago stole genes from oceanic bacteria. These bacteria help to digest seaweed—a major component of sushi and other Japanese cuisines.

Bacteria, from the creepy, crawly, and slimy spaces in this world, may be just as important for making human individuals as brains or genes. Who I am depends on who they are, and vice versa.

THE BACTERIA THAT STICK

We acquire our normal floras (our individual collections of bacteria) with no effort whatsoever. We eat, we breathe, we poke our fingers into the soft parts of this world. In the process, we gather billions upon billions of bacteria—as easily as a ship's hull gathers barnacles.

In the Coit Tower in San Francisco, Victor Arnautoff—inspired by Diego Rivera—painted a mural that depicts a life on the streets of the city. The painting, called *City Life*, is done in muted blues and rusts and tans and in Rivera's Depression-era style. In the background, streets stretch off toward oblivion. In the foreground is daily life in the city.

A police officer is placing a call to headquarters; a man is reading a newspaper; two uniformed sailors are making their way toward the pier; another man is unloading crates of food next to piles of carrots and lettuce; women are holding hands with their children and moving toward the shops; other men and women move between tasks; a man collects the mail while another has his pocket picked as he checks the time; truckloads of grain arrive nearby; an elevated trolley carries people downtown as trainloads of cargo arrive in the rail yards; factories belch smoke; a fire truck races to a blaze; and women stand on the paved walks and share stories. Although it seems chaotic, it intertwines as artfully as a symphony or a colony of ants. Some provide food, some eat, some transport, some collect, and others defend. And each relies on the other, each depends on the rest. Move one, and the sense of what is happening changes.

Inside every one of us, such a scene is played out day after day after day. Intestines deliver foodstuffs, eyes bring the news, the immune system keeps track of the bad guys, the liver cleans the water, and the red blood cells purify the air. The mind worries. The blood flows.

Like *City Life*, each of those inside us relies wholly on each of the others. But unseen in Arnautoff's mural is the true mass of humanity. Inside humans one group stands out from all the rest, outnumbers all the others taken together— bacteria, the ground stuff of life. Underneath and in between every human brick, bacteria thrive. And they mortar it all together.

Our bacteria are not barnacles; they're not just along for the ride. Our bacteria are paying passengers.

And every day we give away a few million and we acquire a few million and—as surely as what we see or hear, as surely as whom we meet or what we eat, as surely as what we wear—those bacteria change us, change what we can do and change who we are. Sometimes, as with syphilis, the change is obvious, other times less so. But there is always change.

My uncle Henry lost his mind because of that—a fling in France, a bacterium—a change. Untold thousands of others have lost their lives because of a single breath of air, a fleabite, a drink of water, a wound, and *Mycobacterium tuberculosis* (TB), *Yersinia pestis* (bubonic plague), *Vibrio cholerae* (cholera), or *Clostridium botulinum* (botulism)—all bacteria.

We notice that. Bacteria reach out and slap us in the face then. It's in the papers, the magazines. It's on TV. It's the reason we thought we went after Saddam Hussein—agents of death and disease, that's what bacteria are. And sometimes that's true. Some bacteria do make us sick and kill us. But that's only part of the story, only the tiniest part.

It's true that bacteria took my uncle Henry's eyes, then his legs, and finally his life. And that's a terrible story. It is impossible to say, of course, just what the bacteria may have given my uncle. But surely they did give him something, surely. And then there's all the rest, the bacteria that gave Henry life to begin with—a considerable gift—and those that sustained him. And it wasn't bacteria that sent Henry off to war. That was Woodrow Wilson. Who knows what was infecting him.

At the end, Beethoven, too, had trillions of companions with him—as he composed, as he slept, as he bathed—constant companions.

Like Henry, among Beethoven's companions was *T. pallidum*, a microscopic spiral-shaped bacterium. In concert with the spirochete and all the others, Beethoven wrote his Ninth and final symphony—the first ever to incorporate the human voice—after he could no longer hear a single note of what he wrote.

According to a study completed in the 1900s, a common feature of tertiary syphilis is auditory hallucinations. And among the most common of these auditory hallucinations are intricate and prolonged pieces of music.

None of us can know for certain which of the notes and phrases, the percussion, and the crescendos (unusual for Beethoven) of the Ninth Symphony were whispered into Beethoven's deaf ears by *T. pallidum*. But they are there, those notes, for certain, they are there. Just as are all the others sung to him by the swarm of life that was Beethoven. All we need do is listen for them.

MIDDLES

Childhood's End

Now that we, from mud and starlight and chromosomes and bacteria, have assembled rudimentary selves, where do we go from here? Only death ends the artistry of self-creation.

As we move from childhood toward another phase, "I"s must often adapt or disappear. The Fisher King's wound, Beauty and the Beast, and perhaps all fairy tales speak of the loss of the child's world and the hard slap of reality. For many reasons, children's eyes see another world, one where anything and everything is real, one filled with chiseled princes and crystalline princesses, benevolent wizards, and one where human eyes, like windows, open onto a complete world. As we teeter on the lip of adulthood, new cracks appear and old ones widen. The reality created for us crumbles and the new "I" must find a way beyond the lies.

The sudden sensation of sex challenges us. When I ask people to tell me something about themselves, nearly always each begins with a statement about his or her sex. Likewise, each of our lives begins with a statement about sex. "It's a boy" or "it's a girl," but the real meaning of that first statement usually doesn't strike us until puberty. Then, we have sexes, and suddenly all the whisperings of televisions and movies, of magazines and mothers and fathers mean something. Most of the time, we yield to those whisperings. They tell us what to wear and what to say. They tell us what to expect from one another. They tell us who we are, men or women, and that there is nothing in between.

The reality is that stories about sex, like stories about every other human thing, are rotten with human wishes and human imaginings. In truth, the simplicity we wish for isn't to be found in the ways of sex. For many that is a hard truth to swallow and it can shake the roots of "I." For others, it is just another beginning.

At childhood's end we expect the good wizards always to appear when we need them. Truth is, wizards are rarely good. And the wizards of "I" can do great harm. Throughout our childhoods and beyond, the world tells us that only the weak fall ill. Human health is the noblest and commonest way of life. Of course, that isn't true, but we wish it was. So we live as though it was true. Until something, like a car wreck, snaps us out of that silvery mist. Then we must seek the wizard. What we hope the wizard will gives us is healing and a return to our old "I"s. Deep down, we suspect that isn't possible. But we know that the wizard will do his best and expect benevolence and the focus of his heart and mind. My wife and I discovered otherwise.

Regardless, we still have the world itself and the truth of that world. As long as we've had thoughts, we've known that the things we see and hear and taste and touch and smell are real and all of the real that there is. Those five senses have given each of us big pieces of our "I"s.

Regardless, we are nearly senseless. The crisp clear edges of a child's reality give way to the blurriness of an adult world. But not just in the sense that our world holds more grays than black-and-whites. Worse.

Most of what happens in our lives completely escapes our senses, and not just for lack of attention. Things we cannot sense, like motives, darken. Things we sense aren't what they seem. And if we look at this world through the lenses of science, we find a disturbing truth. The world we perceive is not the real world. Elephants are guided by songs we cannot hear and bees follow paths we cannot see. The movements of mountains and the meanders of canyons are not for us, nor are the motions of continents or the mysteries of the Milky Way. From the mind of a child come the dreams and fears of the blind.

Aphrodite, the mother of Hermaphroditus, caressing a swan

5

The Opposite of Sex

WITH LISA MAY STEVENS

The simple declaration of "boy" or "girl" at childbirth
sets the final course for most children's lives. And if you asked
someone to tell you about his or her self—or anyone else's self for
that matter—sex always comes up near the beginning. But, in spite
of what we have been told about human and animal sex, every year,
worldwide, more than 65,000 human babies aren't boys or girls
and there are whole species of animals born with no sex at all.

A couple of months into our electronic relationship, Lisa May Stevens sent
me some pictures of herself. In one of these photos, she wore a black gown,
showed quite a bit of leg, and looked like a southern belle—strawberry
blonde, about five feet ten inches tall, hazel eyed. For all the world, like
a southern belle. She isn't though. She's an hermaphrodite from Idaho.

And Lisa May Stevens is a friend of mine.

We met about eighteen months ago, when I sent her an e-mail. At
the time, I had gotten hold of an idea I couldn't shake, and I needed
her input. At the same time, an idea had taken hold of Lisa May, an idea

DOI: 10.5876/9781607322337.c05

that life might not be worth the trouble anymore. Our meeting was serendipitous. Ever since, we've stayed in touch.

"Hermaphrodite" sticks in a lot of people's throats, or it's pitched by people as a taunt. Though coined by Pliny the Elder to describe humans with characteristics of both sexes, when speaking of human beings, the term *hermaphrodite* gets a little slippery, like an icicle in summer. Hermaphroditic plants, on the other hand, form a firm group of individuals known for their ability to take either or both roles in sexual reproduction. And most hermaphroditic animals, though they rarely self-fertilize, can at least perform either party's role in sex and reproduction. No matter what odd thoughts may have slithered inside of Gaius Plinius Secundus's early Roman amygdala, no human has ever been capable of such contortions or contributions.

For all of these reasons, I don't like calling human beings hermaphrodites. But Lisa May insists on it, at least when it comes to speaking about her. She made that clear from the outset.

Since our first meeting online, Lisa May has twice come to Fort Collins, Colorado, to visit. In person, Lisa May is a presence. Her face carries the marks of her masculine past, as do her hands. But her arms remind me of my mother's arms, and beneath her blouses or dresses, her figure is unmistakably feminine. Her genitalia, she tells me, strike her as male and female all at once. That seems to please her a great deal. In Lisa May's story there is a hero and a heroine, and when you look at her, you see both.

Every time Lisa May and I have gone anywhere together, people notice that. One evening at dinner, it seemed like a little tsunami rose from beneath our table and spread across the room as whispers passed and eyes rose from cups of chai or plates of curry to steal glimpses of Lisa May.

You might think people would admire her. And some do, but most do not. Most people seem to find Lisa May's appearance unsettling. Others become angry, as though Lisa May had committed some sin against humanity. Regardless, everywhere she goes, people notice her.

LISA MAY

I always seem to ride the first wave. I love being out front and the shock-and-awe effect. Of course, that fades as most people see me move, gesture, and speak. But from some, I seem to sense hostility, or maybe fear. I still haven't figured that one out. They look, then look away, and soon look back. They watch my movements, my laugh, my gestures. Then their eyes go from my face down my body. Usually they stop at my legs, since I do know how to use those legs to draw attention away from my face. My mother had longer legs than I do. She could command a room like a general. I know the effect legs have on people. Mother taught me well, and in the last two years I have returned to her ways. I'd rather people see me as sexy than as a dog dressed up.

EXPECTATIONS

Once, taking Lisa May back to her motel, I stopped to pick up a cigar. I asked Lisa May if she wanted to come into the store with me. Along with rows of cellophane-wrapped cigars, cigar stores—almost any time of day—contain a few men, cigar-smoking men, if you know what I mean. To my surprise, she accepted my invitation.

Again, Lisa May made waves. She seemed unaware of her effect on these men. I certainly was not. Nothing offensive, but looks that could have extinguished 1,000 cigars sputtered behind those men's eyes. Unmet expectations? Snips and snails and puppy dog tails; sugar and spice and everything nice; pink and blue; boys and girls; men and women; black and white—expectations born of a certainty about sex that equals our certainty about gravity? Perhaps.

As early as four years of age, most children understand that everything comes in twos, and only in twos—mom and dad, grandma and grandpa, does and bucks, boars and sows, hens and cocks, innies and outies. But that doesn't seem to reflect some inborn sense of the permanency of sex as much as it suggests that by age four we have already heard the idea so many times that, for most of us, it has become fact.

Later, well-meaning teachers tell us that chromosomes do that—push us into one corner or another, then bind us up with iron, and leave us there. Chromosomes are final as flint, they say. Y = male, Y-not = female.

And that provides the rigid grout that cements the scales of our beliefs into place, firmly and finally. Unless we meet someone like Lisa May. Then, as the plates heave against one another, our Earth shifts, and tsunamis are born.

But Lisa May's not the problem. In fact, she's the solution to the problem.

Fishy Sex: Defining Nemo

Through shared catastrophe and intractable time, we and fish have grown old together—man and alewife. But the whole time, fish have been outdoing us.

Twenty-five thousand species of fish have been identified and named, but everyone who studies fish is certain there are thousands more. In all, at least 10^{12} (that's a trillion) individuals on this planet call themselves fish.

Humans weigh in at one species and about 6 billion individuals. A pothole in the road of life compared to the crevasse that fish have cut. In fact, among the vertebrates, nobody outdoes the fish. Ranging in size from the Philippine Island goby (about one-third of an inch long) to the leopard-spotted leviathan of the whale shark (about fifty feet long and weighing several tons), no other vertebrate animals compare to fish—not for numbers, not for sheer variety, and not for sexual creativity.

More (perhaps a lot more) than 100 species and 20 families of fish are hermaphroditic, and here we begin to stretch the limits of what we mean by hermaphroditism, what we mean by male and female, and what we mean by everything in between.

Hermaphroditic fish come in two common forms—simultaneous hermaphrodites and sequential hermaphrodites. Simultaneous hermaphrodites have the nifty gift of two sets of genitalia at all times. Sequential herms, as Lisa May calls them, like to rattle back and forth between the sexes, one morning a vixen, the next a lothario.

Hamlet fish are one- to two-inch-long, gold and yellow fish found mostly in the Caribbean and the Gulf of Mexico. They haunt the rainbow-colored reefs in those warm waters, working like little yellow blimps among the sea rods and fire corals searching for food. When they are not hungry, thoughts of sex often dance like little sugared plums inside their tiny heads.

All hamlet fish have both male and female sex organs all the time. That makes them simultaneous hermaphrodites and, apparently, more than a little randy. But these fish do not fertilize their own eggs. Nothing so banal would suit them. Instead, hamlet fish engage in sexual rituals as varied as the tales of Scheherazade.

First, hamlet fish trysts involve multiple matings that last for as many as three nights. And during all of that time, these fish take turns being the "male" or the "female" partner. So, over the course of a single tryst, each fish takes all imaginable roles in the sex act. For such small fish, their lust is great, not to mention their endurance and their penchant for creativity. And when all the sex finally grinds to a halt, both partners are pregnant.

"Though this be madness, yet there is method in't."

Black sea bass, one- to five-pound fish that spawn from Florida to Cape Cod, on the other hand, cover the range of hermaphroditism. In fact, some hermaphroditic sea bass, do, in effect, mate with themselves, though some intricacy is necessary to overcome what seems like an intractable mechanical barrier.

In the watery light that shivers across the sandy flats of eastern coastal waters, some bass spawn as many as twenty times in a single day. Because sea bass have both ovaries and testes, these animals are, by definition, simultaneous hermaphrodites. By any other standard, these fish defy preconception and sport a given sex as briefly and as quickly as the fluttering frames of a rolling movie film. As they spawn, black sea bass alternate between being egg-laying females and sperm-spouting males, a transformation that takes these fish only about thirty seconds. That act makes for lots more fry and the rest of us black with envy.

Among other sea bass, sex is a bit more constant, but only a bit. With these bass, if two females find themselves in a local sea-fern bar and both have sex on their minds, one of the bass simply transforms herself into a male, complete with a dramatic color change and a big boost in testicular output of sperm. Then the remaining female spawns, and the nascent male covers the eggs with sperm newly born from freshly formed testes. Problem solved.

And that's not the last of the fish tales, not nearly.

For decades, maybe centuries, people have known that lots of fish changed sexes. But it wasn't until 1972 that marine biologists began to figure out what motivated these fish to up and abandon their lives as males or females and sprout the genitalia of the opposite sex. Not surprisingly, it turns out that the whole motivation thing is complicated. Every fish seems to have its own set of rules and reasons for swapping sexes.

Beyond black sea bass, wrasse—two- to four-inch-long fish, striped or saddled with black and yellow pigments—flutter in the tepid currents near coral reefs around the world. Some wrasse make their livings cleaning parasites and scar tissues from other fish. Because of that habit, wrasse also swim in lots of home aquaria around the globe.

Regardless, in aquaria or in the hollows among glittering corals, most wrasse begin their lives as females. But chance graces one or two of these fish with testes. As they grow, wrasse develop complex social structures, and by the time these fish reach adulthood, they live in harems of female fish controlled by a single dominant male wrasse. This "alpha" male, through physical domination and perhaps his chemical presence, forces the females to remain females. The females develop a pecking order, with the "alpha" female running the show among the girls. That might seem job enough for a wrasse, but her greatest moment is yet to come. When the one male wrasse dies, over the course of a few days, the alpha female becomes the alpha male and takes over the harem for himself. From veiled damsel to a bearded sheik in a day or so.

And then there are the clownfish—Nemo and all his family. Because they look like some fanciful child's idea of how fish should look—thumb sized and bright orange with vertical black-and-white stripes—these creatures are extremely popular aquarium fish.

Clownfish spin a similar sexual tale, but one with an opposite twist. Darting among yellow sponges and purple anemones, these fish also assemble themselves into groups made up almost entirely of females. But among clownfish, only the largest female in the harem can mate with the single large alpha male. If the large female clownfish dies, the big male hands in his testes, conjures himself a set of ovaries, and becomes a female. After that, the largest of the young females leaves behind her egg-laying days and acquires a skill with sperm. Among clownfish, the few and the proud begin life as females, swap gonads for the grander life of the leader of the pack, and then—for the greater good—reclaim their ovaries and lay eggs as sweetly as any clownfish that ever graced the sea. A tripartite tryst with a sexual subtext unlike any we humans might have imagined.

No matter how hard we may try to squeeze these fish tales into our human stories, sex (to paraphrase J. B. S. Haldane) remains "not only queerer than we imagine, but queerer than we can imagine."

Lisa May is different—given the chance, she will tell you she's fully aware of that. But just how Lisa came to be different is a remarkable tale. Some of Lisa May's cells have two X chromosomes, others have an X and a Y chromosome.

Some of Lisa May's red blood cells are B, some O. But, as incredible as that may seem, that isn't what makes Lisa May's story remarkable.

What makes Lisa May unique is that Lisa began her life as two people—one a boy, the other a girl. The doctors call Lisa May a chimera and a true hermaphrodite. With Lisa, *true hermaphrodite* means that she has reproductive tissues of both sexes—probably beginning with one testis and one ovotestis (a combination of ovarian and testicular tissue).

Human chimeras arise in several different ways, but few begin like Lisa May. Inside of her mother's womb, Lisa May began as two—two zygotes (the single cells that result from the fusion of egg and sperm). One of the two probably would have become—since it contained an X and a Y chromosome—a bouncing baby boy. The other held two X chromosomes and was destined for girlhood. But before either of their dreams became reality, the two zygotes grabbed hold of one another and fused into a single living thing—part boy, part girl— much like Hermaphroditus (the son of Aphrodite and Hermes) and Salmacis (a fetching nymph) fused by their gods into a single being.

Scientist call Lisa May's beginnings a tetragametic fusion—the product of four fused gametes—two eggs and two sperm. The rest of us call her amazing.

Turtle Sex

Even as they lumber up from the sea and carve their way across moonlit beaches to lay leathery eggs, sea turtles don't seem to have sexes. From a distance—or up close, for that matter—turtle biologists themselves often cannot tell a boy turtle from a girl turtle. If you want to know the sex of a turtle, you have to use histology—that requires taking a piece of the turtle, which neither turtle nor investigator much care for—and subjecting the collected tissue to critical scientific analyses. Only after that can a turtle be pronounced boy or girl. But even then, turtle sex involves a lot of assumptions. And if the weather is unusually cool, a newly hatched turtle may have no sex at all.

Turtles don't even have sex chromosomes, and genes don't seem to play any direct role in deciding whether a turtle will end up as a girl turtle or a boy turtle.

Turtle sex is mysterious.

At the heart of that mystery, lies—like a glowing coal—the temperature of the sand and the sea and the air, the temperature of the pond and the forest and the river. A shift in the wind, the slippery movements of clouds, a storm front, a warming trend, and the sexes of turtles drift—another male, fewer females—one direction or the other, and a turtle's future looks a little pinker or a little bluer.

Somehow out of all that—and the turtle itself, of course—the warmth and the warp of the sea lay down the course of turtledom like a highway. Turtles can do nothing but follow—and that includes sea turtles and tortoises as well as land turtles and tortoises—all marching to a single drummer, the weather.

LISA MAY

Lisa May wasn't born Lisa May. She began life as Steven. But even then, when Steven's father wasn't around, Lisa's mother often called the baby "Lisa." Something of a confusion for her, but only at first. Her mother dressed Lisa May in girl's clothes and talked with her about the ways of women—how much makeup to wear, how to tease, how to stop, and how to please her mother. Lisa May soon figured out the rules and how to be her mother's daughter and her father's son. Reality had little to do with it. Practicality ruled Lisa May's childhood. Her father hated Lisa; her mother could do without Steven. Lisa May did what needed doing.

But once Lisa May left home, reality reared its cyclopic head. The easy move between sexes just didn't work so well anymore. Lisa May needed one sex, a fixed, hard-and-fast sex. She settled as Lisa May for a while. But after a traumatic rape, she reached out to Steven. She took hormones—major doses of testosterone—she bound her breasts as tightly as she could, and she took a job as an ironworker. Steven worked hard and made his way in the world. He met people and made friends. Later, he married twice—both times to women with bisexual tastes. He divorced, he struggled, he tried suicide and failed.

It's not an easy thing to do, move between sexes. I find I did some of this [opting for Steven or for Lisa] as a habit rather than thinking about it. In the end, though (if this really is the end), I remembered how I was taught at a very young age and how Lisa is second nature for me. I just had to let go of Steve. But to let go of something, even when it seems like a ball and chain, is not as easy as one might think. At times, I have fond memories of life as Steve, but as Steve my objectives in life were less clear. With Lisa, I am more focused on things that matter to me, gentler, more accepting of life's ways. But I do have one clear fault as Lisa—I trust people way too much. I will be working on this a lot in the near future, a whole lot.

'GATOR SEX

Between 1948 and 2006, the state of Florida lost nineteen people to alligators. That's nineteen for certain. It seems likely that a few others who stopped showing up for roasted crawdads and banana fritters at the Lost Hope Crab Shack also ended their lives in the arms (and jaws) of a 'gator. 'Gators are of mean temperament and gluttonous appetite. And they are noisy.

A snout, long and big as a suitcase, filled with saw-blade teeth, those vertical slits in the center of eyes that pop out of their heads like the headlights on old sports cars. And then there is that tail, armored and as full of fight as a python. Alligators do evoke something reptilian, something buried inside of humans a long, long time ago.

Alligators hibernate in the winter, stop eating when the temperature drops below 73°F, and make sex while the sun shines. As with turtles, alligator sex has nothing to do with chromosomes. Sex comes to alligators from their surroundings.

Alligators, crocodiles, and caimans don't have sex chromosomes, and inside these creatures, there is no consistent genetic difference between males and females. Instead, the he/she bifurcation fork splits its tines after fertilization. And the road most traveled by the zygotes depends mainly on the weather. Male or female is left to the vagaries of sunlight, water, and wind.

When the mean ambient temperature is between about 88° and 90° Fahrenheit, nearly equal numbers of males and females hatch from American alligator clutches. But when the mean temperature falls by as little as two degrees, American alligators stop producing any males. The same thing happens when the mean temperature rises by about three degrees. Crocodiles and caimans appear to have similar pacts with the weather.

In the end, just what makes an alligator or a crocodile or a caiman a he or a she isn't clear. It might be hormones produced by the hypothalamus, it might be something else. Whatever it is, it watches the skies and the sands with the eyes of a prophet, waiting for just the right push from a star's light.

LIZARD SEX

Lastly, there are lizards—another scale-plated, prehistoric-looking, bug-eyed link to our past. Lizard sex is a little easier, and safer, to study than that of alligators and crocodiles. So in studies of temperature-dependent sex determination, lizards have been slightly more popular as research subjects. All three families of geckoes do it. Some iguanas do it. So do a lot of other lizards, but not all. Sex as a warm hand on a cold body.

SEX IN THE SUN

For a raft of animals, sunlight and sex are inseparable. Whether many creatures on this Earth become males or females is purely a matter of where the mercury falls along the length of its glass tube. In the laboratory, the sexes of some frogs will even reverse when the temperature is raised or lowered. Whether that happens in the wild isn't clear, but it seems likely. Soil temperature, pond and ocean and river and rock and swamp temperatures—driven by the light of a star almost 100 million miles away—make turtle and crocodile sex, push lizards onto lifelong paths, flip fish from male to female, and make alligators fat with hormones.

Some argue that 'gator sex is a prehistoric idea since gone south. But lizards evolved much more recently than alligators, and lizards far outnumber big amphibians likes 'gators and crocs. So the sexual ambiguity of lizards can't be so easily tossed off as some ancient aberration.

Are we really who we seem to be, or have we been misled by millennia of the misinformed?

A turtle's sex defines nothing. Turtles, all turtles, are just turtles. Sex is a costume worn to deceive the bloodied eye of time. A sham to make more turtles.

Fish fake sex simply to entice a lover and 'gators have no opposing sexes, just sex. Lizards wait for a fickle photon to plant their seed; and frogs, regardless of their initial bent, find what's necessary to further frogdom.

No one of these creatures is first of all (or even last of all) a boy or a girl, a bull or a cow, Mars or Venus. And no one of these creatures would ever imagine themselves opposite of or at war with the other sex. Because half the time they are the other sex, or something in between, or might have been if the sun had simply shown for another hour or two on the day he or she was born.

For me, at the gnarled root of that tree lies one tough acorn.

Teachers and textbooks, clergy and color TV all remind us that sex is as fixed and firm as the iron hoop of a human chromosome. Only two possibilities exist, and those two are as different from one another as night is from day. But try telling that to a turtle or skunked-eyed 'gator who has seen starlight turn an amorphous chunk of protoplasm into a boy or a girl or something else. Try telling that to a sea bass about to fertilize her own eggs or a saddle-backed wrasse at that slippery point between egg layer and sperm sprayer.

LISA MAY

In 2006, Steven and Lisa May parted company once again. Nothing about Steve felt right to Lisa anymore. She quit the testosterone and began a well-designed and supervised regimen of estrogen. Her breasts swelled, her voiced crept up the scale, her mind quieted. Then Lisa May fell in with a group of people whose lives were more like hers. None of them, of course, had a past or set of chromosomes that could match Lisa May's, but they thought differently about sex—how you got one and what you might do with it—differently from most of the rest of us. And that appealed to Lisa May. One of Lisa's new family had her breasts removed as the first stop on her road to manhood. Lisa May admired that, too.

Sex change makes some people in this world edgy, others angry—it puts a crack in the whole sex-as-concrete thing. But it doesn't bother Lisa. In fact she encouraged her friend to treat his sex like he might his mind. If after all these years your thoughts or your sex just don't make any sense any longer, perhaps it's time to change them.

I have learned to walk though deep water without losing a step, to flick my hair to one side or the other, and of course how to smile the smile that stops traffic. When others stare at me, I just look past that and smile at each one. That breaks the ice. I often spend way too much time even food shopping. But who cares?

Wood louse

6

Lousy Sex

Among the most successful and numerous of all the
animals (the arthropods) sex is an even curiouser thing. For these
animals male-versus-female sex has more to do with infections
than it does with genes or chromosomes or temperatures.

This morning, I am escorting a wood louse out of my kitchen and onto
the lawn. It is early spring. The air is warm and full of promise, and
as I launch the balled-up creature lawnward, my thoughts turn to sex.
Unusual sex, mysterious sex, infectious sex.

Humans will use almost any excuse to think about sex. But wood lice
will never be voted among the top ten reasons for this. Wood lice—roly-
poly bugs, pill bugs, potato bugs, sow bugs—are those armor-plated
crustaceans that scatter, on fourteen jointed legs, like cockroaches
when the light hits them or, when we poke at them, curl themselves
into impenetrable scale-covered balls.

Sort of creepy. But it gets even creepier. Wood lice harbor one of
the deepest and darkest tales ever told about animal sex. Irresistible,
uncontrollable, unforgettable tales of sexual exploitation.

DOI: 10.5876/9781607322337.c06

73

Sex, it turns out, isn't always about who you are or where you've come from. Among wood lice, sex has a lot more to do with where you're headed for and what you've got—or, more accurately, what's got you.

Nearly 3,500 different species of wood lice carpet our planet with their scaly hides. And—crawling through our gardens or the moist soil lining our homes' foundations and breathing through gills—wood lice are hiding something, something sexy.

If you spend time among the detritus that has collected in the dark and damp spots around your yard, you're bound to encounter wood lice—they're everywhere you'd just as soon not be. And if you spend a lot of time poking through the leaf mush and garden goop, you might notice that besides their slightly prehistoric appearance, there is something truly weird about wood louse sex.

Surprisingly, it's pretty easy to find out if a wood louse is a girl or a boy. According to *Invertebrate Anatomy Online* here's all you need:

Dissecting microscope

8-cm culture dish

Living or preserved terrestrial isopods

Chloroform

Cotton

Applicator stick

When you nudge the living isopod (wood louse), the creature will reflexively curl up in to a ball. A short while after you drop the bug into the culture dish, it will uncurl, almost always on its back. This is where things get exciting. If you want the louse to hold still, you'll have to resort to the chloroform-soaked cotton. Personally, I like them squirming. Now just slip the culture dish under the microscope and take a close look at the critter's pleon—the last seven plates at the rear on the underside.

The long slender endopods of the first two pleopods of males are modified to serve as copulatory organs, or gonopods. The same is not true of females.

It's just that simple.

I am not certain what the applicator stick is for. Maybe it just makes the whole thing seem a little less prurient, a little more clinical.

After you have done this a dozen or so times, it gets really easy. Also, after you have done this a dozen or so times, you will begin to notice something odd. Every louse you picked up has turned out to be a female louse. Beginners may doubt their technique at this point. They shouldn't.

Most species of wood lice are all or nearly all females. The few species of wood lice that do produce males produce very few compared to females. For many humans, this discovery is disconcerting. We take for granted that the nature of evolutionary forces, reproductive advantage, and old-fashioned sexual predilections will always lead to male:female ratios of about 1:1. Wood lice take nearly nothing for granted, and among wood lice, sex ratios never naturally approximate equality. We also take it for granted that whether a child of ours becomes a boy or a girl is a result of something we as parents did (that whole chromosome thing). But among wood lice, sex is completely out of their control.

As it turns out, sex is also beyond the control of most insects; many spiders, mites, and ticks; a fair number of parasitic worms; a few shrimp; and some lobsters. For all of these animals, sex has nothing to do with any of the things we think of as essential to sex.

Wood lice, if they could, would love to go along with the rest of us, spread their chromosomes far and wide and lay down equal numbers of boy and girl eggs. But this world rarely gives wood lice such an opportunity. If allowed, wood lice—like birds—produce males that have two Z chromosomes and females with one Z and one W chromosome. And left to their own devices, wood lice produce about equal numbers of lousettes with ZZ and ZW chromosomes—nearly equal numbers of males and females.

Trouble is, wood lice aren't allowed.

Instead of a chromosomal grip on the tiller of their own sexes, the world has given wood lice an infection. A very small thing, something that people overlooked for years, usually takes control of wood lice sex before chromosomes even get a shot at it. So the lack of "long slender endopods" that you observe under your microscope was no mistake. Male wood lice are a rare commodity.

Until the 1970s, this skewing of wood lice sex ratios remained completely mysterious. About then, a microbiologist gone rogue decided to take a close look at wood lice eggs. What he found there forever changed the way we think about sex, if not lice.

Inside wood louse eggs, this microbiologist found fistfuls of bacteria. Not a common thing with eggs. And one particular type of bacterium predominated—a bacterium called *Wolbachia*. In spite of the fact that no one had ever seen anything like this with insect eggs, these bacteria didn't stir up much interest among microbiologists or entomologists. A few years later, though, another miscreant microbiologist looked to see what would happen if he treated wood lice with antibiotics to kill the bacteria.

Amazingly, wood lice treated with antibiotics began producing about equal numbers of ZZ males and ZW females. Whatever was wiping out wood louse males, antibiotics cured it. Among wood lice, sex was a disease. Even worse, an infectious disease.

This didn't fit with anybody's model of all-female societies or, for that matter, with anyone's idea of rollickingly good sex.

To deal with their disappointment, and to cover the evidence that males might not matter so very much, microbiologists explained all of this as a rare example of a bacterium somehow skewing the sex ratios of one species of wood lice—one of those "gee whiz" sorts of things, but of no real importance to understanding entomology, let alone sexual reproduction. After all, wood lice are, well, lice. How important could any of this be?

That explanation didn't hold up long. When scientists started looking elsewhere for *Wolbachia*, it turned out that most species of wood lice, most species of insects, many shrimp, some lobsters, some spiders, and a lot of little mites all have an infection, an infection that makes most of them into little girl lice or shrimp or spiders or mites.

THE LOW ROAD TO DOMINATION

The more places microbiologists looked for *Wolbachia*, the more they found it. And when the investigative dust settled, it was apparent that *Wolbachia* is the most common infectious bacterium on Earth. And far and away the world's champion sexist.

Hardly an evolutionary blip. Rather, the interaction between *Wolbachia* and its host is one of the most, if not the most, frequent symbiotic relationship(s) in all of biology. Like people and cell phones, almost every bug has a constant companion.

Wolbachia infects about 65 percent of all insects. Insects are by far the most numerous of all multicellular animals—about 10^{15} individuals and as many as 30 million different species. That means that about 6.5×10^{14} animals (equivalent to 100,000 times the number of human beings on Earth) and at least 1 million species of animals are all infected with *Wolbachia*—each and every one making its living only at the whims of a bacterium. And then there are the infected spiders, the mites, the ticks, a fair number of parasitic worms, more than a few shrimp, and several lobsters all equally infected and equally prone to the tricks of bacteria.

That means that *Wolbachia* infects most of the animals on this planet. Which in turn means that, among us animals, most of us have absolutely no say in whether we end up in the pink or the blue crib. Bacteria determine sex for most of us. So while we call this an abnormality, an infection, and a disease, in truth it is simply the way of life for most of us. Regardless of what Sister Irene may have told us in high school, it's microbes, not chromosomes, that hand out sexes to animals.

Just when the first bacterium took over the reins of a wood louse's sex life can't even be surmised. But how so many animals continue to be infected is not so inaccessible. *Wolbachia* live, among other places, inside the eggs of all of its hosts. So every time a baby louse, mite, wasp, lobster, and so on is born, it is already infected with *Wolbachia*. That explains how this bacterium continues to infect so many animals. But why do most of these animals end up as females?

Wolbachia, it turns out, is an obligate intracellular parasite—it can only make a living inside of someone else's cell. *Wolbachia* doesn't have everything it needs to make more *Wolbachia*. What it lacks, someone else must provide. So to keep their families going, each of these bacteria have to find a room (cell) of their own inside of which they will make baby *Wolbachia*.

Outside of cells, *Wolbachia* die. That presents another problem. How to get from louse to louse? For *Wolbachia* to move from one animal to another, it must move inside of a cell. Lobsters and shrimp and wood lice don't transfer cells from one to another, except during mating.

That's where *Wolbachia* saw its opportunity. If it could find its way into sperm or egg, the bacterium was on the high road to a new generation and a bright future. Here was an obvious solution.

But only eggs have enough cytoplasm (the semi-liquid stuff that surrounds the nucleus) to accommodate *Wolbachia*. Sperm are just little outboard motors attached to DNA. Because of that, sperm have no room for bacteria. So the future of *Wolbachia* lay with the egg.

Solving the egg problem, though, wasn't good enough. If sperm and egg came together and made a male wood louse, it was a dead end for *Wolbachia*. That's because any bacteria inside of those cells would be squeezed out as the new male's spermatogonia got small enough to become sperm. That meant that at least half the time, *Wolbachia*'s journey would end in the black pit of the belly of a sperm. Evolutionarily, if not personally, that was a problem.

MINIMIZING MALES

To *Wolbachia,* the solution seemed obvious—limit the number of males. And the bacterium immediately set about developing an array of ways to do that. One of the first tricks *Wolbachia* put to use was a process called cytoplasmic incompatibility.

Without *Wolbachia,* most matings between male and female wood lice are productive. But among infected species, when a *Wolbachia*-infected female mates with an uninfected male, nothing happens. No boy wood lice, no girl wood lice, nobody—cytoplasmic incompatibility. *Wolbachia* do that. Only when an infected female mates with an infected male—providing the same species of *Wolbachia* infects both sexes—does a multitude of new tiny sow bugs come into this world. Somehow, *Wolbachia* see to it that if one parent is uninfected, Dad's chromosomes disappear as the embryo is just getting started. And that means that new lice appear only when the conditions are just right, just right to produce one more litter of blue-bonneted lousettes.

And that's good. But not good enough.

The Power of Parthenogenesis; or, "L" is for Louse

To improve its own odds of survival, in some species of wood lice and wasps, *Wolbachia* has simply blown the males clear out of the picture. In these wasps and wood lice, *Wolbachia* turns all developing babies into ZW baby girls. On top of that, in these critters, *Wolbachia* infection confers the power of parthenogenesis on each new female. That means that in these species every female can produce more baby lice or baby wasps without the need for a male partner. Parthenogenesis is females making more females, making more females, and so on. Babies, lots of them, and no need for sex.

In these wood lice and wasps, not only has an infection changed the sex of the young, it has also driven these animals from the ranks of the sexual reproducers and has returned them to their past as asexual beings producing only female young. No need for any males.

Treatment with antibiotics will force some female wasps and wood lice to revert to sexual reproduction and hatch male and female animals that have to find one another to make more like themselves. But in other wasps, *Wolbachia* has found a way to close all avenues to the past, even those that might have been reached through antibiotics.

Asobara tabida Nees is a relatively common wasp that lays its eggs in fruit-fly larvae—little worms that hatch from fly eggs. As the young wasps develop, they slowly consume the living fly larvae and emerge as fledgling wasps, already infected with three species of *Wolbachia*. Though antibiotics cure the sexes of some other infected species of insects, treatment of *A. tabida* with antibiotics makes the females sterile. In fact, elimination of *Wolbachia* causes these wasps to stop making eggs completely. It seems that on top of everything else, *Wolbachia* has also taken control of oogenesis (egg making) in this species. So if the bacterium cannot have its females and its future, then neither will the wasps have theirs.

From a simple bacterial infection, this relationship has evolved to a fully dependent symbiosis between wasp and bacterium. Together, these individuals have reached a point where neither can live without the other. And of course, that relationship is absolutely dependent on the wasps' continued production of females, which they happily provide. Infectious sex, infectious egg laying. What was once two has become one.

Wolbachia has also established essential long-term relationships with several filarial nematode worms, including *Brugia malayi* and *Wuchereria bancrofti* (the worms that cause elephantiasis), and *Onchocerca volvulus* and *ochengi* (the worms that cause river blindness in places like Guatemala).

Elephantiasis (aka lymphatic filariasis) is a terrible parasitic disease with enormous swelling of one or more limbs as a result of infection and blockage of a person's lymphatic vessels. River blindness (onchocerciasis) is the second leading infectious cause of blindness in the world (trachoma is number one). River blindness also results from infection with a filarial worm. Damage to the eyes is only one of many, many problems caused by *O. volvulus*. Surprisingly, antibiotics kill the worms that cause both of these diseases.

But it isn't because the antibiotics kill the worms themselves. As is the case with all other worms, antibiotic compounds have no direct effect on these worms' physiology. But the worms have developed a complete dependence on an infection with *Wolbachia* for their survival. On top of that, *O. volvulus* obtained from the blood of infected people treated with doxycycline (a powerful antibiotic) could no longer cause river blindness. It appears that the human interaction with the *Wolbachia* inside the O. *volvulus* worms is an essential piece in the puzzle of river blindness. No bacteria, no river blindness.

Infectious femininity, infectious blindness, macabre symbiosis.

MALE KILLERS

Two-spotted lady beetles are sometimes mistaken for ladybugs, an easy mistake to make with these red- and black-spotted insects. That misidentification leads people, for no apparent reason, to treat these beetles much better than people treat most beetles. Perhaps we would be less likely to treat these beetles so kindly if we realized that two-spotted lady bugs, even as they crawl across our palms, are systematically killing all of their male eggs—a seemingly heinous act.

In truth, though, the beetles can't help themselves.

Most two-spotted lady beetles, along with *Acraea encedon*—beautiful orange and black butterflies—produce only female offspring with normal female sex chromosomes. Antibiotic treatment causes both species to begin to produce genetically normal male and female beetles and butterflies in nearly equal numbers. That suggests that most of the time something bad is happening to about half of the eggs belonging to these butterflies and beetles. In other words, it seems very likely that all of these female insects are producing equal numbers of male and female embryos, but the male embryos just don't make it through to hatch as live animals.

Inside these bugs and beetles, the ever-resourceful *Wolbachia* has found yet another way to limit the number of males. In some beetles and butterflies, *Wolbachia* simply kills all the male eggs. And it turns out that two-spotted lady beetles and *A. encedon* butterflies are just two examples. *Wolbachia* drives many other insects to murder their male eggs.

By the way, among male-killing species of bugs—as they hatch, the females dine on the dead eggs of their would-be brothers. So among the male killers, nothing goes to waste, except for the occasional male ego.

MALE CHANGERS

And still that wasn't enough for *Wolbachia*. More ways of eliminating males had to be found if the bacterium was to achieve its full potential.

In some species of insects, *Wolbachia* simply changes genetic males into females. As I said before, without infection most insects produce ZZ males and ZW females in nearly equal numbers. But if you examine some *Wolbachia*-infected insects, you find that—even though the bugs continue to produce equal numbers of ZZ and ZW animals—all of the offspring are females.

Even when faced with the daunting task of overcoming genetic predispositions, *Wolbachia* is unfazed. It seems that *Wolbachia* has found some way to manipulate hormonal control during male development. As a result, even though the baby bug has male genes, *Wolbachia* overcomes that predilection by several different and generally poorly understood mechanisms.

One trick some species of *Wolbachia* use is to close the spigot on male androgens and force the developing bugs on the female railroad. If these were human beings, we'd call them pseudohermaphrodites and label them as biological abnormalities. Among these insects, pseudohermaphroditism is the norm. And if antibiotics or some other killer destroys the infecting *Wolbachia* during fetal development, the baby bugs are all intersex.

Male, female, or in between: all under the control of a bacterium. All of it the result of an old but persistent infection. In the end, though, the result is a whole population where chromosomal differences don't correlate with anyone's sex. It doesn't matter whether an animal is ZZ or ZW. They are all females, and all infected with *Wolbachia*. In spite of our fondness for stories of chromosomal supremacy, our insects suggest that genes, by themselves, don't mean squat.

BEYOND *WOLBACHIA*

If that weren't enough to subvert your ideas about sex, it turns out that *Wolbachia* isn't the only bacterium that can change an animal's sex, not even close. The false spider mite *Brevipalpus phoenicis* is a tribute to the force of life, the force of life and the power of compaction, that is. These mighty mites are tiny—about 3/250 of an inch long by 16/2,500 of an inch wide, which is about as wide as a human hair and about one-third the size of the largest bacterium ever discovered.

Tiny.

False spider mites infest mostly plants where they can do a lot of damage, especially to citrus plants. I assume they are called false spider mites because they do look a little like spiders, but they aren't, they're mites, another sort of arachnid.

When entomologists first studied the false spider mite, the researchers were amazed to find that all of these animals were haploid females. Most animal cells are diploid—meaning they have an even number of chromosomes, half of which came to them from each parent. Normally only sperm and egg are haploid and capable of performing their functions with only half the normal number of chromosomes.

False spider mites were the first animals ever found to be uniformly haploid—meaning that each cell inside of each mite has only one copy of each chromosome and only one copy of each gene. That shook the foundations of a couple of aspects of evolutionary biology.

Of course, maintaining a strict haploid existence precludes even the occasional sexual tryst. So, a long time back, false spider mites took up the practice of parthenogenesis, producing children without sex—all of them females. Among these mites, apparently alone among all the animals, there are no chromosomal differences between gametes and somites—no differences between the reproductive cells and all the rest of the cells of the body.

That drove the evolutionary biologists to probe further. How could this have possibly happened? How could an entire species of what must have once been diploid animals be reduced to these hapless, haploid remnants? Actually three remnants, since there are three closely related groups of false spider mites that are all haploid.

False spider mites' eggs seemed like a good place to start trying to answer those questions. What those biologists found inside false spider mite eggs opened another door everyone imagined shut for millennia. False spider mites' eggs were full of bacteria. Not *Wolbachia*, surprisingly, not even a distant relative of *Wolbachia*, but a whole different phylum of bacterium—*Cytophaga-Flavobacterium-Bacteroides*. Full of bacteria.

Thousands of these bacteria filled the cytoplasm of every false spider mite egg the scientists examined. Out of curiosity, the scientists treated some of these eggs with antibiotics. For comparison, they also hatched out some eggs that had seen no antibiotics. Infected eggs all developed into females, while "cured" eggs all developed into males.

Even though every false spider mite today has only one copy of each chromosome and only one copy of each gene, these experiments gave clear evidence that once upon a time, these mites had been sexual reproducers. Once male and female false spider mites had walked proudly across the fields and the fruits of this Earth. But no longer. A strain of *Bacteroides* (another bacterium) changed all of that. Who knows how long ago the last natural male false spider mite fell prey to an infection, an infection that slowly, or maybe not so slowly, changed the course of false spider mite history forever. Gone was sexual reproduction, gone were homologous pairs of chromosomes, gone was romance as we know it, gone were male false spider mites—all gone, gone forever because of an infection. And today, without that infection, all of the false spider mites would disappear within a single generation.

THE PROBLEM WITH THE PAST

In the classroom, most of us learned three things about sex as we made our way through our science courses. First, since everybody—or at least everybody who is anybody to us humans—uses sex to reproduce—there must be some enormous evolutionary advantage to this sexual reproduction thing. Second, males and females always arise in about equal numbers so everybody gets an equal shot at reproducing. The same must be true for everyone else. And last, the thing with its hand most firmly on the throttle of gender control is a chromosome.

The textbooks and the teachers we learned from, though they meant well, were wrong.

The vast majority of animals in this world have dramatically skewed sex ratios, numbers nowhere near the supposed ideal of 1:1. The majority of the animals in this world are females—females who have given up on sex, but whose species still flourish. For most of us, chromosomes have nothing to do with sex. Instead, it's all about bacteria, mindless bacteria bent on their own success.

Sex for most of us is, in fact, an emerging infectious disease. Eggs full of bacteria, bacteria whose sole interest is making more bacteria. To them we are nothing more than a vehicle, a vehicle designed to carry them from place to place and to move them from egg to egg. We persist only because we can deliver these bacteria into our children and the world beyond. Our past may have been filled with sterile misunderstandings about sex, but our future is septic.

———

In nearly slow motion, the wood louse rolls through the air toward the newly green lawn. Now, I am reminded of Strauss's *Blue Danube* and Stanley Kubrick's epic film *2001: A Space Odyssey*. The scene where the shuttle from Earth and the space station lock into in a slow minuet.

The wood louse hits the lawn, bounces upward once, and falls back. For a few seconds, it just lies there—feigning death. Then slowly, the isopod uncurls it chitinous body and rights itself.

Something like David Bowman's last words rolls through my brain. In Arthur C. Clark's book version of *2001*, Bowman—the sole surviving crew member on the deep-space probe *Discovery One*—at last arrives near Saturn and finds a monolithic stone left in orbit there by some other intelligence. David dons his space suit and, inside one of the extravehicular activity pods, jettisons himself from the giant ship *Discovery*.

The music inside my head switches to that of another Strauss—Richard—and *Also Sprach Zarathustra*.

As he nears the monolith, David's ragged breaths rasp like bellows through the telecommunications device inside his space suit. Finally, directly above the anomaly near Saturn, from the oval window of his EVA pod, he stares into the monolith.

"The thing's hollow," David says. "It goes on forever—and—oh my God! It's full of bacteria!"

An illustration showing Sputnik 2 and Laika

7

The Wizards of "I"

Much of what we come to think of our selves, how we see our
selves, and what we believe our selves to be capable of come to us from
others. How we are treated, how we are spoken to, things we are told
as truths, and whether we are loved are the bolts and rivets of the constructed human self. Without art, science and medicine may become
the wrenches and the chisels that will undo those bolts and those rivets.

November 1957. It's cold, flesh-cracking cold. The barren steppes roll
off into the icy haze, unmarked save for a few low clumps of dead grass
and three or four brick buildings. Light like flint flickers over the plain
and a polar wind slams at the bricks over and over. Nothing changes. It
seems nothing here ever changes, ever has changed.

As if in argument, the center of this land of sleet and ice begins to
burn deep orange. Miles off, the light gathers itself and roasts the thin
air. Kerosene and liquid oxygen come together as though they have
been waiting for this moment for a million years. A gray hare raises its
nose to the wind, turns to see the light.

DOI: 10.5876/9781607322337:c07

At the flame and the noise, men inside one of the small blockhouses run to windows filled with smoked glass. As they watch, the fire burns stronger and brighter. Seconds pass—nothing more happens. Though no word has been spoken, each of the men inside is afraid that nothing more will happen. Finally—a millimeter at a time—the fire begins to rise. Slowly, so slowly it seems certain to fail, the flame pushes itself and a two-stage rocket from the frozen ground at Baikonur Cosmodrome, USSR. The hare lowers its ears and its eyes fill up with flame.

Behind the blistered bricks, the men turn to one another and smile. Overhead, the light grows dimmer and dimmer, then disappears altogether. *Sputnik 2* has left this world. Two stages, 10 tons of metal, 253 tons of kerosene and liquid oxygen, 1,118 pounds of aluminum nose cone, and 13 pounds of living flesh.

November 2000. It is a mild dry day. Just beyond Pueblo, Colorado, my wife and I are headed south toward New Mexico for Thanksgiving. The light has gathered over the mountains to the west. A cool wind is blowing down their shoulders and everything smells of sage and an early winter. Colorado slides quickly past our windows. We speak of times past and times to come. As we talk, I am scanning a road map, hoping to pinpoint our whereabouts—an odd and nearly useless habit of mine.

Suddenly, something appears in the road before us. Gina swerves to the left to avoid it. I look up from the map I've been studying and see that we've driven partway into the median, at about seventy-five miles per hour. Weeds are whipping past, slapping the front then bottom of the car. Gina swerves back onto the roadway. The car twitches oddly to the left, and we are abruptly in the right lane of the southbound highway. Gina reaches again for control of the 4,000 pounds of steel and vinyl beneath us. She tries hard to bring the car back in line with the road. Again it twitches. Again she reaches.

Time itself splits open. I have no past. There is only future.

Slowly, the car heels over onto its side—Gina's side. Sound returns like a fist. The windows shatter. Everything we'd packed for the trip is in motion inside the car. Gathering speed, the car quickly flips three, four times. The world around us loops past our windows. Only we are fixed, all else is motion. Again the silence. A computer hovers behind me, in free fall. A pen is rotating just past my fingertips in the center of nothing. Gina and I are still, pinned to the instant by the nylon straps of our seat belts.

Suddenly, the spinning stops. Gravity slaps a final time at us and our Explorer comes to rest upside down in the dirt in the median. The wheels spin, the car sighs, yellow dust pours in through the shattered windows. All that was with us is strewn across 100 yards of short-grass prairie. Everything is red or yellow or black. I reach to my side and fumble with the seat belt and finally release it. I drop back into time and land on my head in the now-foreshortened cab of the Explorer. I reach through the broken window and pull myself out into the oddly warm dirt, the strangely yellow and green grass. I do not know if my wife is dead, alive, in this state, on this planet. It is quiet again.

I see Gina in Pueblo twice after that. The first time we accidentally pass in the hall, each of us strapped to a white-sheeted gurney. Each of us pushed by green-clad people with severe expressions hanging like bats from their faces. We hold hands for a moment. We squeeze each other's hands and then we are wheeled off in different directions. The last time I see her there, she is paralyzed by medication and stuffed full of plastic tubes. She can't speak or move. I squeeze her hand. She stares at the tiled ceiling and the empty face of the fluorescent tubes. Then she is gone in a clatter of helicopter blades and rush of cold air.

"Will she die?" I ask as they wheel me back into the rings of the CT scanner.

"She might," the doctor responds. "But if she survives the first forty-eight hours, she may make it. Her bruised lungs are her greatest enemy just now," he says solemnly. He is wrong.

Forty-eight hours later, I catch up with Gina in Denver. Pieces of her are missing. Gina—her right hand strapped to her bed, a thick vinyl tube pushing air into her lungs, her legs swaddled like two swollen baby Jesuses, and her left arm taped and strapped tightly against her chest—looks to me like God himself has beaten her, repeatedly. Her face is a topograph of bruises. A piece of her intestine is gone and her belly is split like a burst purple grape. I can see inside of my wife.

Gina's bed crouches among a cluster of such beds, arranged like the petals on a daisy—black and blue, with plastic hoses stuck in each petal. Over each body, a bank of lights burns surgically. Skin-colored tile stretches over the concrete beneath and dribbles off toward the waiting room. Darkness seeps from the walls. Everything happens slowly. People, in pale green, move from place to place, following some choreography I can't fathom. Nothing changes. Soiled dressings, spilt urine, clotted blood. Isopropyl alcohol. The smells of hell.

For the first time I seriously consider the possibility that we didn't survive at all. The reality that this might be our punishment.

I sit. The people around me change faces, but they keep telling the same stories. I understand that the stories are terribly important. In spite of my stupor, I understand that, and I try to listen. But the stories are too complicated.

The first two days in the ICU, Gina imagined herself blind. The darkness was that complete. In reality, the pounding she received bruised her face so badly that both eyes were swollen tightly shut. The whispers, the people in transit, the noise, the smell. Hieronymus Bosch vitalized. The only lights that flickered were those behind her eyes, flickered over a landscape no one else has seen.

I was the first to tell Gina about her eyes.

Instead of giving her back her sight, the huddled doctors divided her up with the knives of their whispers and argued among themselves about who got the first shot at her. There was a "back guy," a "shoulder guy," a "knee guy," and a "lung guy." The lung guy more or less had his way for the first two days after Gina arrived. The bruises and the pneumothoraces demanded that.

Today, the "back guy" wants a crack at my wife.

Five of Gina's vertebrae fractured when our car flipped. The back guy worries that if her back isn't fixed as quickly as possible, the broken bones might damage Gina's spinal cord. If that happens, she might be paralyzed for real. Probably he's right. I have no way of knowing. The night all of this happens, I am in Fort

Collins—trying to shake some of the foggy-headedness and the hallucinations that are chasing me. The doctor calls to ask if it is okay with me if he splits Gina's back. From his tone, it is obvious to me that he thinks I, not Gina's sister, should be there in Denver. While trying to make sense of what he's asked, I am watching a broad-leaf plant turn into a large, black dog with slobbering lips and a curled pink tongue. Eyes like roasted chestnuts. I tell the doctor to go ahead.

He hangs up. I lay back and wait for the drugs to take effect.

That night, the back guy cracks the skin above Gina's spine, lays in two metal plates, and drives ten stainless steel screws, big around as pencils, into her spine with a power drill. Then he opens up her hip, carves pieces of some bone there, and packs the bone chips between her bolted vertebrae. Gina's hip will hurt worse than her back for a long time to come.

When I meet the back guy in Gina's room a couple of days later, he tells me he wants to show me a picture of my wife. Apparently, he has forgiven me for my absence or, more likely, forgotten. He sticks a black chunk of film into the clips of a light box. The picture shows nothing but my wife's lower spine and the steel he has added to her busted bones. He seems very proud of the steel. He seems very pleased with his photo of my wife.

Gina and I hold hands when we can. But we never speak to one another. We can't speak to one another because of the ventilator tube pushing Gina's lungs open and closed. The tube blocks her larynx. She can say nothing. Most of the time, there are only the two of us and the wheeze of the machine breathing life into her, then sucking it back out of her. One morning, Gina slides a piece of paper into my hand.

"I need to get out of here. They're trying to kill me," is scribbled across the blue paper.

How can I argue with her?

———————

The engineers gave Laika air enough to last for six or seven days. *Sputnik 2* would likely orbit Earth for six or seven months. On Earth, time was short. In space, time would be no factor. The Americans had made fun of the Russians' first satellite—it weighed little and held only two tiny transmitters. The United States assured the Russians that U.S. scientists weren't wasting their time on anything so trivial. That made everyone in Russia angry, but it made the Soviet premier especially angry. Nikita Khrushchev himself insisted on a new satellite—larger and with something alive on board—to be launched within one month. They would show the Americans.

Orders were orders. People began throwing things together. A launch vehicle was assembled from an existing R-7 ICBM—once targeted, perhaps, at the United States. A payload capsule was welded together. A Tral-D telemetry system was bolted in to transmit data back to the Earth during fifteen minutes of each orbit. Two spectrophotometers were added to measure solar radiation (both ultraviolet and X-ray) and cosmic rays. And finally a television camera was mounted in the "passenger" compartment. The camera would transmit 100-line video frames at ten frames per second. Then it was time to find a "passenger."

Two men were sent to Moscow. At a local shop, they bought a kilo of meat and walked back onto the city's streets. In a nearby alley, they enticed a dog—part terrier, part Samoyed, all cur—into their van. There they fed her the rest of the meat and chained her to the van for the trip back to Baikonur.

Now that they knew the size of the animal they would hurl into space, the men began to fashion a harness for the dog. When that was done, they modified the passenger compartment to fit the mutt, added a generator to provide oxygen for a few days, and a plastic bottle that would hold some water. An air cooler was welded in along with a food container, not very large. Recovery of the satellite and the dog was never even considered. Not nearly enough time to plan something so complex. There was time only for blasting the dog and her machine into space and into the faces of the Americans. As though she knew, the dog barked constantly. So the men at Baikonur named her Laika—Russian for "barker."

Pictures were taken so all the world might know. Laika lay, looking almost happy, inside her capsule. The harness wasn't in place, of course, and the lid of capsule was off. They gave her lots of food for the pictures. But not at other times. Her weight was still of utmost importance.

The ascent from Baikonur Cosmodrome is more than Laika thinks she can stand. She is crushed to the floor, her ears pinned to her head, her jaw slammed shut by the flat slap of gravity. Fear like she has never known floods her mind as the ship races toward the heavens. Saliva runs between her lips, through her teeth, onto the floor of the capsule. She urinates.

Inside the twelve-feet-high by six-feet-wide cone, Laika understands none of what is happening, or why the men in Moscow grabbed her while she tried to eat what they seemed to offer so freely, then swept her off to Baikonur and packed her into this capsule. At the moment, though, the roar and the ragged rattling of steel against steel have stripped her of all curiosity. She has space only for now.

The shoulder guy gets the next shot at Gina.

From the very first, the man who does the work does not believe Gina's shoulder can be fixed. He tells me that before the surgery even began. In the waiting room, at the top of the stairs. Beneath us, a hundred people sit in poorly padded chairs waiting for a turn to speak with a nurse or "visit" with a doctor. A hundred people afraid of their own bodies. A hundred people with no one else to turn to. A hundred people from every nook and cranny in Denver. And above them all, we speak of the fate of a single woman's shoulder, and it means everything.

"I'll try to repair her shoulder," he says matter-of-factly, "but I may have to replace it. If that happens, then I don't know. I'm less enthusiastic about that outcome."

"Why?" I ask.

"Prostheses just don't work very well in people your wife's age. No one really understands it. She's too young for this surgery."

"Oh."

But something has to be done. Time is running out.

I sit back down with my sister, her husband, and Gina's sister. I try to tell them what the surgeon told me. None of us understands. Among the soiled fabrics in the waiting room, we sit with others equally confused and wait.

The surgeon reminds me of his pessimism again immediately after the surgery. We meet this time in the surgery waiting area—a mildly purple and gray space where eight or ten of us sit with heads bowed while people we barely know work our loved ones over with knives. The lights hum overhead. People come and go. Men and women in green scrubs arrive periodically, looking like caterpillars, and speak softly to one group or another in the waiting room. The people who hear either cry or smile. Then, usually, they leave, holding one another's hands.

Gina's surgeon, short, dark haired, in his green scrubs, Italian, looks at me through eyes brown as my mother's and tells me to give up. "I tried," he says, "to save her shoulder. But there just wasn't enough good bone left to hold the screws. Her bones are really not very good."

Her fault. She should have seen it coming, done more.

"So we replaced her shoulder."

With what? I want to ask, but don't.

"But, like I told you, I'm not very optimistic about how that will turn out. Not at all. She's too young. These things don't go very well in people like your wife."

"What does that mean, actually?" I ask.

"I don't expect she will regain much use of her shoulder."

And he shows me, lifting his limp-wristed hand to elbow level, his upper arm tight against his body. "Something about like that is average."

"How is she?"

"She's doing just fine."

What does that mean? I want to ask, but again I don't. I am afraid this time he might show me.

Instead, I make up my mind that he is wrong. Averages made no difference. Gina hasn't ever been average. Making up my mind, though, makes no difference—to him, to Gina, or to what was once her shoulder.

As he walks away, I can't shake the image of Gina's shoulder lying in a stainless-steel garbage can somewhere.

He could be wrong, though, that doctor. He could be.

As it reaches for the blackness above this Earth, Laika's ship groans and cracks like a breaking wrist. The brutal forces of sound and gravity just keep lashing at her. The air inside smells of ozone. Laika whimpers, then tries to bark, but the noise and the gravity swallow her words. She can see nothing. She cannot even open her eyes. Laika believes this is the end. But after what seems hours, the hammering stops and the horrible weight lifts itself from her back. She rises from the floor until the straps of her harness pull her to a stop. Frightened but released from gravity, she draws long ragged drafts of the stale air that pollutes the capsule.

Two hundred miles away, men huddle together to listen, watch—taking notes, laughing, congratulating one another. Some hope to learn from this mission, others hope only to slap the Americans' faces. One way or the other, learning isn't an essential part of this project. The essential piece is to be sure that the Americans know that Laika is up there.

The Americans know all right. But not about Laika. Not yet.

Laika knows nothing. None of those who spy on her ever speaks to her. She knows only the strange blackness of space. When the Americans discover Laika, they call her "Muttnik."

As Laika enters Earth's orbit, the second stage of the booster rocket fails to separate completely from the capsule where she lies. So, now the capsule orbits booster down. That is, the booster remains pointed at the Earth, and the capsule window faces the stars. Because of the attached second stage, Laika's tomb is easily seen by people all over the Earth. A star itself, spinning through the empty heavens. We watch.

A tiny window had been cut into the lid of Laika's capsule so she could see as well. But because of the way the rocket booster and payload orbit, Laika never sees the Earth again. She sees only stars—icy needles burning in a black dome. Laika has never seen so many stars. In the photos of her taken from the video camera mounted near the window above her head, Laika lies in her harness with her forelegs crossed and stares out the window. The light of the camera momentarily blinds her. For 15 minutes out of every 103, the stars vanish and the capsule fills with light. For 15 minutes out of every 103, Laika imagines someone has come to get her. Of course, no one has.

––––––––––

The gauges continue to pulse, the air remains chemical, and once sunlight, undiluted sunlight, falls through the window at the foot of Gina's bed. But even that isn't enough to lift the weight from her chest, or to lift the wooden fear from me. It is ten days before anyone tells me that Gina will likely live. I never asked. No one volunteered.

The night they drew straws for Gina's parts, the knee guy drew the short straw, so he has to go last. Only after the intestine was shortened, after the shoulder was gone, after the back was bolted, after most of what could be done to Gina had been done to her does he get a crack at her knees.

But before that happens, Gina insists we return to Fort Collins. She is well enough now to leave the ICU, and she is sick nearly to death of the hospital in Denver. The lung guy has finally managed to wean Gina off the respirator, and she is ready to leave. The knee guy in Denver is very disappointed that we are leaving. He really has developed a little love for Gina's knees. He offers to do the surgery at 5:00 a.m. on Saturday before we leave. Gina refuses. He reluctantly recommends a knee guy in Fort Collins, and we leave for home. But not together.

She leaves in an ambulance. I leave with my sister and her husband. It is December now, but the sun is shining. After all the darkness, the sun is shining. That, of course, will change.

A day or so after we arrive at the hospital in Fort Collins, we meet another back guy. He has been chosen for us by someone we do not yet know to follow up on the work begun by the back guy in Denver. From the moment we meet, this back guy makes it clear he has little interest in and even less time for following up on another back guy's work. Gina's back is not her own. Her back belongs to a nameless surgeon in Denver, though she must bear it. From a distance, the new guy looks at Gina's back, suggests some new X-rays, and, without once touching Gina, he leaves.

The next day, Gina and I meet the new shoulder guy. He shows up an hour or so late one evening, still wearing his ski clothes. Tall. Solidly built. Darkly handsome. He holds her chart while he sits in a green plastic chair. Legs crossed. Shoulders slack. Absorbed in his thoughts as water is absorbed by cotton underwear.

"Doctor, I've had no physical therapy for my shoulder. None. I'm worried about that." Gina.

"People like you don't rehab well from surgery like this. Oddly enough, you're too young." Doctor.

"You know, I really wish you wouldn't begin by pigeonholing me with averages. All I want is to work as hard as I can to bring my shoulder as far as I can. Let's not have any preconceived notions." Gina.

"I'll order the physical therapy." Doctor, shrugging.

The end.

He never touched her shoulder and he never ordered the physical therapy. "People like Gina" don't rehab well. He just wanted out of the room.

Once during the next three weeks, a therapist (apparently on her own) offers to help Gina. She shows Gina how to do things Gina didn't think she could do. And she shows her exercises that might help Gina's arm. Gina is so appreciative she cries and holds the technician's hand for as long as she can. A month later, we insist that we see the shoulder guy again. He agrees to meet with us on Christmas Eve.

He doesn't come. Later we learn he changed his mind at the last minute and went skiing with his family. The room distorts around Gina's bed. The white sheets wrinkle and redden. The pale green walls fold in on themselves. Gina would not get her shoulder back, he figured, no matter how much attention he gave it. So he gave it none.

The hallways lengthen and narrow. The overhead lights turn yellow. I look at other people now with the eyes of a hare—fearful, jealous. The darkness comes again.

Gina's shoulder then does something no one had ever imagined it would. It begins making bone. At first that seems good. But then it doesn't stop. Her shoulder goes on making bone. And not just shoulder bones, not just scapula and humerus. It begins making bone everywhere. "Heterotopic ossification."

Bone grows over the top of her shoulder and into the plastic and steel of her prosthesis. Bone grows down her deltoid and into her trapezius. A hard milky cap forms almost overnight and seals forever beneath its opalescence the handsome young doctor's predictions. But it isn't, as he had forecast, her youth that took her arm; it was calcium and cartilage and confusion that robbed her. Something no one thought could happen.

When we finally meet with the shoulder guy again, we confront the surgeon with his absence. He resents our confrontation. "These things don't work well in people of your age. They just don't," he spits at us, watching as Gina struggles to lift her left arm.

"I never asked you for any sort of guarantee," Gina mumbles.

"That's not the point. These things don't work well in people your age. We all knew that!"

"We never asked for a guarantee that this would work out," Gina shouts.

Finally, me to him: "You never even said, 'I'll do my best to see to it that things turn out as well as possible.' You never once said you were sorry about what had happened and you would do all you could."

Pause.

Pause.

"I'm sorry," he says.

Me: "I guess we'll find ourselves another physician."

"I understand," he says.

He understands nothing.

Near the end, Laika understands. Everyone has abandoned her. That is clear now. No one will come with the light. But why? She remembers little of the days at Baikonur. She barked a lot, of course. That's was only because she was frightened at first. That doesn't seem so terrible now. But here she is alone, her own filth clotted in her fur. The stink of urine, and the heat. It must have been something she did. How else could she explain this? And it must have been something so terrible, so unforgivable that the men who had been so kind to her, who had fed her and played with her, no longer wished even to speak to her, to touch her, to be anywhere near her. It hurt her to imagine she had done something that so angered these men.

Something that bad, it seems, she should remember. And she tries. But even when she tries very hard, she can't recall what it was. All of them had seemed so nice, so anxious to make her feel important. How could she have been so stupid?

"I'd be friggin' depressed too."

Gina's out of the hospital now and these words are from the new (third) shoulder guy. This guy isn't really a shoulder guy; he's sort of a knee guy. Only he doesn't do knees like Gina's. He was the best the hospital could find after we fired shoulder guy number two. Number three is explaining to Gina why she should take antidepressants. To encourage her, he is explaining that he certainly would take antidepressants if he were anywhere near as fucked up as she is. This is what is known as bedside manner.

We're at a clinic in Fort Collins. Gina has struggled onto an examining table. I'm sitting in a chair on lifts, little blocks of wood under the chair's legs. I don't know what the lifts are for. The doctor is wearing his lab coat. I'm thinking I should have worn mine. The overhead light is thinking nothing at all, I guess. A bottle full of tongue depressors and a box of Kleenex watch us from the countertop.

"I've known lots of people who got along just fine with a completely fused shoulder. *(It isn't your arm that's the problem. It's your head. Get over it.)* That's what you should do. I feel good about this already. You need an antidepressant."

Gina isn't so certain. After all, it is her arm, regardless of what she has been told. "What about options for my shoulder?" she says, tearing up.

"I'd be afraid I'd make it worse. People your age don't do well with these prostheses, and I might make it worse. I think you're just going to have to accept that this is the way it's going to be for someone your age."

"What about rehab?"

"If it hurts, I wouldn't do it. Your shoulder is not going to get any better. But I'll talk to a colleague. A shoulder guy I know in New York. Get his thoughts about all of this."

Of course he won't. I've become cynical.

Gina, though, seems a little encouraged.

What shoulder guy number three does do, regularly, is pass whatever information we give him onto the other shoulder guy—number two, the one we fired. Semi–shoulder guy number three would then simply pass back to us whatever originally fired–shoulder guy number two told him about Gina's shoulder. Neither number two or number three ever touched her shoulder.

After all, it wasn't their shoulder to begin with. Just whose shoulder it is, no one knows for certain. But it is time for Gina to get used to it. Everyone agrees on that.

"Take the antidepressant, and maybe you should try some Oxycodone as well," shoulder guy number two soothes now. "I'll present your case at rounds next Friday to see what all the specialists here think about this."

He won't.

"But I don't have much hope for surgery. I think the antidepressant is the answer. I feel good about that. I'll leave a prescription at the front desk."

None of these is even his idea. All of it—the opinions, the prescriptions, the pain killers, the antidepressants, the gloomy forecasts, and all the shitty advice— is still coming to us from the man we told to get out of our lives. This man, the one in front of us, is just the go-between. And the advice we are given is still filled with the wisdom of averages and the commitment of the bored.

Gina begins to cry seriously. The doctor walks out. She has rehearsed her story for two weeks before this meeting. The story she will tell them about who she is and how hard she will work, how proud she can make them if they will just give her a chance. But no one ever asks her to tell her story. No one ever gives her a chance. No one. Now there are only her tears—rainwater up from the sea.

Indifference has ensnared us within its tepid arms. Gina's shoulder is forfeit. Next it will be her knees.

The air is nearly gone now, Laika knows that. Each breath is harder to draw and hold. She can't stop panting. It is so hot here. Not yet seven days in orbit, and her support systems are failing. The coolers are no longer working properly, and the oxygen generator is going as well. She has never been so frightened in her life. The men on Earth know what's happening to Laika, but they cannot tell her. Besides, the scientists have accomplished what they wished to anyway, even in so few days. Days full of night. Days full of stars. Days full of failure for Laika, who still doesn't know what she did wrong. There is no need for more.

The last of the oxygen bleeds into the cabin and into Laika. She settles as best she can in her harness and the weightlessness of her capsule. Confusion pushes every other thought from her mind. The burning in her lungs slows and finally stops. Unforgiven, her heart simply quits beating.

Six months later, Laika's corpse burns to black ash when her ship reenters the Earth's upper atmosphere, slams against the fierce wall of air the dog once breathed so unthinkingly. Fire and ice. A streak of flame cracks the darkness over the Northern Hemisphere as she hurtles across the sky. Then there is only silence and darkness and ash.

The light scatters. We gather it again, focus the rays with the lenses left to us. Images rise. Nothing is as it was.

Shortly after I finished this essay, at the 2002 World Space Conference, Dimitri Malashenkov revealed that the first accounts of Laika's life and death had been twisted to accommodate the needs of Mother Russia. In truth, Laika was dead within hours of liftoff—a victim of her own blind panic and the slow roast of the sweltering cabin. During the launch, Laika's heart rate tripled, but then trickled downward once the rockets shut down and she slipped into a perpetual free fall. Still, her heart took three times longer than usual to return to near-normal rhythms—her terror was so great. Then, the cooling systems began to fail; they were an afterthought anyway.

Dimitri explained how he and the other Sputnik 2 scientists had listened as Laika's heart rate shot up once more. Each time the capsule came into range of Russian telemetry, the dog's pulse flared in the men's ears, then vanished, flared then vanished, then simply vanished. Rhythms beyond numbers, grabbing lungfuls of ever-hotter air, less than seven hours into her space flight the dog's heart gave out.

But the capsule spun on, shouting to the Americans of Russian supremacy. Announcement of Laika's suffering and death would only divert attention from the true significance of what was happening here. So Dimitri and the others told no one.

Until October 2002 at the World Space Conference in Houston, Texas. Three dogs had been captured and groomed for the Sputnik 2 mission: Albina, Laika, and Mushka. Albina, testing early rockets, was twice lofted into the cobalt dome over the Russian steppes. Mushka was used to develop and evaluate life support and telemetry. Laika, of course, was hurled to her death high above this sweet old world.

In a BBC article reporting on Malashenkov's revelations, Dr. David White-house (an astrophysicist and online science correspondent for the BBC) concluded only, "Despite surviving for just a few hours, Laika's place in space history is assured and the information she provided proved that a living organism could tolerate a long time in weightlessness and paved the way for humans in space."

No doubt, for Laika six hours seemed a very long time. As for the future of space travel, who can say?

What was once unbroken short-grass prairie stubbles out gray-green from here to the horizon. Only the foreground of this landscape marks the century—twin black strips, dashed white and the constant thrum of tractor-trailer rigs. From the Shell station at the Colorado City exit on I-25, we can see just where the Ford Explorer came to rest on its roof. A tiny strip of median marked with bits of cardboard and papery bunchgrass worried by a dry wind.

We've been back here maybe a half-dozen times. Usually we stop, hearts thumping. So small a space it seems, too small to have saved us from spinning into the northbound lanes of traffic, too small to have held us in its gritty palm until rescuers arrived and pushed us back through the membrane into our crusted futures. Too small.

As we pull away form the gas pump, we stop for a moment longer, eyeing the ragged spot where our lives split in half. But, like Baikonur, the place speaks only of hard winters and human manipulation.

Gina reaches with her right hand for the indicator wand on the left side of the steering column, her left arm clenched permanently to her side. As she pushes the lever skyward a small yellow beacon flashes from the right front of our car, ratcheting out a warning for all the others. We glance back at the median, then move out into the cataract of the freeway.

Double rainbow

8

Dreams of the Blind

Facts and truths keep us whole and sane.
But nothing in this world is at it seems, not the light, not
the dark, not the sound of an elephant weeping, nor the touch of a
lover's hand. And most certainly not the truth.

"Victor Delgado!" a woman in a cage shouts into a tired PA system.
"Front!" The speakers rattle with her efforts.

She looks like she's spent the better part of her life on a barstool,
smoking cigarettes, drinking watered gin, and waiting on her next ex-
husband-to-be. Hard and polished like the runners on an old sled. I'm
glad it's not my name she's calling.

I turn my back to her and scan the men standing in this room with
me to see if I can guess which one is Victor. The room is not big—
wooden floor and the far wall lined with thin, tall windows. A couple
of sofas sit slump-shouldered against the near wall, and six or eight men
are scattered like stones across the dirty floor. Some of them are talking,
others are watching me. Every one of them looks like he knows about

DOI: 10.5876/9781607322337:c08

me already—without a word from me. Some sort of pissant that has never spent a day in jail. Maybe they're right about the pissant part, but they're wrong about the rest of it.

I turn toward the woman in the cage. She's looking at a magazine now. Splashed across the page that has drawn her away from me, Cher appears in the arms of aliens, all caught in the blinding explosion of a photographer's flash.

And then Victor comes walking out from a side hallway. I turn to watch. The man is shorter than I expected, muscular with a rolling gait like he might have been a sailor, and thick through the shoulders. His black hair has thinned across the top of his head, and his goatee is mostly gray. He's older than I expected. He has time in jail written all over him.

"Hello," he says to me. Which, unexpectedly startles me. I say nothing to him.

"Are you Gerald?" he asks, like he might ask a child.

"Hello," I say. "Yes, I am."

He reaches for my right hand, his palm slick and dry and his grip to the point. We shake. He smells of lemony cologne.

"Do you want to sit outside?" he asks, looking me in the eye.

It is July. The air is thick with summer, and a little breeze is rolling down from the mountains.

"Sure," I say, in spite of myself, and then add, "if it's all right," eyeing the lizard woman and her tabloid.

"Yeah, we're allowed outside," Victor says, glancing at the woman with the magazine. She steps sideways to spit into a trashcan. Something about that makes me think a little better of her.

While my mouth fills with its own spit, Victor pushes me aside and heads for the door. He is wearing a plain white T-shirt and blue jeans, no belt. Old tattoos on both his forearms have been poorly scratched into scars. His neck stands up thick and creased across the back, deep brown.

As we walk, I stare at his back, muscled, filling up his shirt, and I watch the slow roll of his hips. His walk says more to me in ten seconds than his words will for four months, or so I think.

Outside, a small knot of men stands on the concrete patio, idly smoking cigarettes and watching the college girls across the street. Most of the men are young, younger even than Victor, whom I'd put at about forty. No one speaks to us.

We walk around a corner of the building, and Victor points to a spot on the lawn near the wall. Both of us sit. Victor leans back against the red brick and drops his hands across his knees. His fingers fall across his blue jeans like polished wooden pegs. For a moment, I worry about his crime, and how it will be for me working with him.

We are sitting against the wall of a halfway house for convicted felons. A stack of marrow red bricks pierced by narrow windows. Someone has cut the grass short, but the cracks in the walks are full of weeds. Colorado State University is just across the street to the south and to the east is a sorority house. Some of the women are sunning themselves on their balcony. Even I can see they know they have an audience.

Victor pushes his wispy black hair back from his forehead and looks at the campus and buildings to the south. "You work there?" he asks.

"Yeah."

"What do you teach?"

Someone has told him what I do. I wish they hadn't. I hesitate.

"I teach stuff about diseases mostly, human and animal diseases. And I teach some writing classes."

"Not like this one, I'll bet."

"No, not like this one."

"That okay with you?"

"What?" I ask.

"Teaching about diseases."

"I like it okay."

Then there is a lull.

"When did you quit school?" I finally ask, simply to say something.

"About third grade, I think," Victor says. "We moved around a lot."

My head starts to spin once more, a carousel of thoughts. I most want to know what Victor did to land him in this place. I don't have the nerve to ask. Some things I do know, though. Victor isn't here because he failed to pay a parking ticket. And Victor is bilingual—he speaks both Mexican Spanish and American English with ease and fluency. But he cannot read or write a word of either. That much they told me when I took the job.

"Yeah, it's true," he confides. "I can't read neither."

Curiously, his tone seems one of shame, but his movements suggest a certain sort of indifference. The contrast registers with me at some subconscious level, but I leave it there to ferment.

Because I have nothing else to offer, I show Victor the books I was given to help with our studies—adult literacy books, not Dick-and-Jane stuff. Victor acts interested.

We settle against the wall, uncomfortable with our new roles. For a while Victor talks about himself. He was born in Eagle Pass, Texas—a smallish border town south and a little east of Del Rio. His mother is still alive and lives here in town. She is Mexican, or once was. Victor isn't sure about his father. Thirty or so years ago, Victor's father came up missing. So Victor's mother packed up her son and went off in search of a living. They bounced from place to place until they landed in Fort Collins. She found a regular job at a Wal-Mart here, and because of that bit of money, they stayed. But by that time, Victor was too old, by a long shot, to go back to grade school. So instead of school, he went out and found himself a job, sometimes several jobs. The jobs paid okay, and as long as he had money, no one asked what he did for that money. But a few years later, Victor ran afoul of the law. Probably more than once.

We talk a little about the books and the lesson plans—I'm making it up as I go—and we talk a little more about Victor's past. Victor's dark-brown eyes are quick, and I can see he reads a few words, but only a few. After about thirty minutes, we quit for the day. Enough for now.

I like him, I think. And I'm pleased with this beginning. I walk to the curb and watch the evening sky moving toward a deeper blue.

BLINDNESS

I am fascinated by blindness. The finality of it, the black face of it, the long, hard night of it.

Partly, maybe mostly, fear gave birth to that fascination—a deep, ice-pick-in-the-eye sort of fear of unabated darkness. Because of blindness, I am fascinated by light, the stuff the sun rains down on us all day long.

Children nearly always color the sun yellow. Sunlight may be a little yellowed by its passage through the Earth's atmosphere, but the light that streams from the sun—as nuclear fusion hammers hydrogens into heliums—is white, white-hot and blinding, white as a freshly bleached sheet of cotton. But if you push it through a prism, the white light splits into all the colors of the rainbow—violet to red—splayed out like a harlequin's fingers. All the colors we can see lie buried beneath the whiteness of sunlight. That in itself would be enough. If sunlight never did anything more than that, more than hold and hide the fistful of colors we can see, that would be enough for me.

But it is not enough for the sun.

On either side of the prism's rainbow our eyes see only darkness. But if you lay a thermometer in the darkness below the rainbow's red or above its violet, the mercury inside that glass tube will begin to twitch. Better yet, run your finger along the whole length of the prism and see that when the light dies at either end, the heat doesn't. There is something more here, something we cannot see.

Science and a chunk of glass gave us the rainbow, a considerable gift. But with that gift comes a staggering story of loss—blindness and loss.

Though we've at least four other senses, seeing is truly believing. Ask a friend to tell you about his bedroom, his mother, himself, and he will first tell you what he saw—not how the bed sheets felt nor how his mother smelled, not how his own voice sounds to him, or how his tongue tastes inside of his mouth. When ask to describe our world, nearly always, we first speak of light. Human stories are mostly stories about the things we have seen, stories about light. So powerful is sight, it is the only sense we must eliminate for sleep.

A touch is a touch, but sight is the world, our world.

Even some people who are demonstrably blind will speak to anyone who might listen of all the things they "see." This condition is called Anton-Babinski syndrome, or Anton's Blindness. Named after Gabriel Anton and Joseph Babinski, it is one form of anosognosia—lack of awareness of some physical defect.

Theodore Kaczynski, the Unabomber, had anosognosia. Facing a first-degree murder conviction, he refused to let his lawyers push an insanity plea. Ted said he wasn't sick—he had a grudge, for sure, but he wasn't sick. He was willing to die to prove that.

To prove they can see, people with Anton-Babinski syndrome confabulate and make up stories to hide their blindness. They describe people who aren't there. They propose to walk freely across a room and then stumble over the furniture because they don't see it. They create tales about what they themselves are wearing and see colors and spaces where none exists. They will describe in detail the ties around their physicians' necks when there are none.

To the rest of us, that seems insane, indefensible, mad.

But the invisible rainbow that leaks from the end of the prism's knife blade hints at a general madness among humans, a madness about light and sight.

VICTOR

August.

I ask Victor if he drives.

"Sure," he says.

"How'd you get your license?"

"They read you the questions."

"Oh. I didn't know you could do that."

"Yeah, you can, man. All you gotta do is ask."

"You'd rather not ask, I guess."

"Yeah. I'm gettin' too old to be asking people to read stuff to me."

He smiles at me, at least his mouth smiles at me.

I smile back. "I guess maybe that's why I am here, huh?" I suggest.

"I guess so," he replies, watching the girls across the street.

I tell Victor to open his book to the first lesson.

"My girlfriend can read, too. That's another reason I gotta learn to read. As long as I don't read, she thinks I'm dumb."

"Does she live around here?"

"Yeah. Just a couple of miles over that way." Victor points vaguely toward the northeast.

"How much longer do you have to be here?" I ask him.

"One year, maybe a little more, maybe a little less." Victor scratches the side of his head as he speaks, leans back against the wall of the building and sighs. "They gave me eight years. I served five and a half." He pauses, then asks, "You ever been in jail, man?"

"Yeah," I say, and Victor looks surprised, maybe doubtful.

"Really?" he asks.

"Yes," I admit.

I notice Victor's book is still closed.

"What for?"

"I called a cop an asshole."

Victor smiles at me. "Not too smart, huh??"

"Not too smart. What are you in for?" I ask reflexively, then wish I hadn't.

Victor pauses for a moment, runs his tongue over his lips. Then he says, "It was cocaine, man. Cocaine." He says nothing for a second, like he can still taste it. Then turns toward me and says, "They caught me with seventy kilos coming down from Cheyenne."

"Wow, seventy kilos, that's one hell of a lot of cocaine. How'd they catch you?"

"I was speeding, man. You know, you always tell yourself to drive real slow, man, real slow, and don't break no laws. But when you got that much stuff in your trunk, all you want is to get rid of it. So you get in a hurry, and you forget all the things you said you weren't going to do."

More than anything, I'm relieved. I had imagined murder, maybe even a gruesome murder: the body chopped into thumb-sized pieces and stuffed down the garbage disposal. Or maybe aggravated assault—beating someone just nearly to death. Poison, putrefaction, pederasty, or knives, almost anything to do with knives. Cocaine? Hell, I'm okay with cocaine.

I point him toward the book once more.

"Why'd you call that cop an asshole?" he asks.

Light

Light is carried by photons and photons move a little like waves in the ocean. If you watch ocean waves roll onto the beach, you see the waves break onto the beach at semi-regular intervals. Those intervals are the result of the spaces between the waves. That space or distance from wave top to wave top is the wavelength. With ocean waves, those distances usually can be measured in feet. Light waves are a little more varied.

We can see light with wavelengths from about 400 to about 750 nanometers. A nanometer is one-billionth of meter. A meter is about a yard long. Tiny. The shorter the wavelength of a photon, the higher its energy. So visible light is composed of photons of very high energy and very tiny wavelengths.

Sunlight contains photons with wavelengths from less than 10^{-16} to greater than 10^8 meters—from unimaginably tiny wavelengths to a wavelength longer than 62,000 miles. That is an enormous range. At best, you and I see only around .000000000000000001 percent of that light.

In truth, things are even worse than that. Human eyes are far from 100 percent efficient at converting light into sight. The rods and cones (visual receptors) in our eyes don't even face the right direction for maximum reception. For that and other reasons, our eyes convert less than 10 percent of incident visible light into vision. So my blindness is at least ten times worse even than the abominable percentage I already calculated.

Most of what happens right in front of our eyes goes completely unnoticed.

In fact, mathematically, the difference between what the sighted and the blind see is insignificant.

VICTOR

September.

"You vote, Victor?"

"Nah, I ain't allowed. But my mom, she does."

I can't believe I so easily forgot that convicted felons cannot vote. Stupid of me to bring it up. I change the subject. "What's your mom like, Victor?"

"She's pretty good to me. Her life hasn't been so easy. She works at Wal-Mart and doesn't have too much time for anything else. She wrote to me a lot, though. And all the time I was in prison, sometimes I wrote to her. And I used to make her these little . . . like, retablos."

"What did you make for her?"

"Uh, these little pictures—religious stuff. I made them from matchsticks and aluminum foil. I glued the matchsticks to the aluminum foil in different shapes so they looked like Jesus or something. I got really good at it, man. I made the Virgin Mary once, too. All from nothing but matchsticks and aluminum foil. Then I'd frame them and send them to her. She liked them a lot. I made some for my girlfriend, too. I'll show you one sometime."

"I'd like that," I say. "Where did you get the matches?"

"Oh, everybody collected matchsticks for me. I just used the burnt ones."

"They let you have matches?"

"Some of us," he says.

"How long did it take you to make those pictures?"

"A long time, man. A month, maybe. Long time. But there was lots of time down there in Cañon City, lots of time." He looked away. "I'll try to get my mom to bring me one of those pictures so I can show you."

"I'd like that."

"Before I sent the pictures to my mom, I wrote poetry on the back of them. She really liked that."

Finally, I begin to listen to what I am being told. "Wait a minute, you can't read or write."

"Oh, yeah. A guy I knew down there took care of that for me. I told him what I wanted to say in the poems, and he'd type up these little cards on his typewriter. I taped them to the back of the matchstick pictures. Guy in my cell read the letters to me."

"How'd you come up with ideas for your poems?"

"Oh, I'd remember things, like about how one of the places we lived looked like in the moonlight. And I'd write something about that."

"Did your girlfriend like the poems, too?"

"Yeah, she said she liked them a lot."

Week after week, I arrive at the halfway house and, as long as the weather holds, Victor and I sit outside and work on his assignments. Sort of. I still carry the books they gave me when I signed up for this job, because that is all they gave me when I signed up for this job. I have no idea what I am doing here. I know nothing about teaching adults to read. In fact, I have no idea about teaching anybody how to read, adult or otherwise.

Victor seems fully aware of that. And to my surprise, he doesn't seem at all bothered by the fact that I am completely horrible at this. Instead, he seems amused and more than a little pleased.

Summer is folding up shop. The air smells of apples and students have begun to litter the university campus. Victor and I have made limited progress. I'm dancing around him like some freshman at the senior prom and he, well, he's just smiling and watching me, black eyes aglitter.

I go back to the place where I first volunteered for this job and ask them for help. I explain how things are going. I tell them about how nervous and frustrated and disappointed I am. I explain how I know nothing about what I am doing, nothing. I remind them I am working with a dangerous criminal.

They give me more copies of the two books they gave me to begin with. "Maybe it will help to give Victor his own copies of these books. He can write in them if he wants. Perhaps then he will feel more involved."

The next time we get together, I give Victor his own copies of the two books, and then I give him assignments for the coming weeks. He places small X marks next to the assignments in his book. But he doesn't immediately seem more involved.

Each week the assignments go mostly undone. But this isn't like my courses at the university. This is a convicted felon I'm dealing with, and I am not about to lecture Victor on the value of his homework and my time. Maybe it was only possession of cocaine, but it was a lot of cocaine and for the past five years plus, Victor has been living with people who did some really bad things to other people—butchering, bombing, and the like. Some of that is bound to rub off on a person. So mostly, I just work with whatever Victor brings me.

The weeks pass and we claw our way into phonetic pronunciations—which don't make a whole lot of sense, even to me.

I struggle to give him useful rules for picking the words off the page and spitting them back at me. We make a little progress. Victor can now read simple sentences. Though in my more cynical moments, I believe Victor always could read simple sentences, and he is only now giving this to me because he feels he must give me something.

Tired of fighting it, I flip the literacy book shut. There is a steady murmur from the men smoking near the front porch and a low rumble coming from the traffic moving along Laurel Street between us and the campus. "Maybe that's enough for tonight, Victor."

"Okay."

"I'll see you next week."

Victor heads toward the door.

"Hey, Skeeters," a man calls out. It is one of the men smoking on the cement patio near the front door. The other men turn toward Victor, but as Victor turns to face them, all but one drop their gaze toward the stones at their feet. With their eyes, he and Victor hold one another for a moment, and then Victor turns toward the door.

Skeeters?

DOING TIME

Because of Victor and blindness and light, I've become interested in time.

Beyond Fort Collins to the west, the foothills of the Rocky Mountains rise into the afternoon sky. Those mountains are actually the second set of rocky mountains. The first set disappeared a few hundred million years ago. The ones I'm looking at are a bit younger.

A little more than 100 miles beyond those mountains, the Colorado Plateau begins its slow march into Arizona. The plateau itself is an unusual uplift of prehistoric stone carved into tortured shapes by eons of water and wind. At the very heart of the Colorado Plateau lies the Grand Canyon—a mile deep, and in places nearly twenty times that wide. The Grand Canyon formed as the Colorado Plateau, along with a few other things, rotated around a point somewhere in northern Texas. About 20 million years ago, as the plateau began to spin on its molten base, it rose—about a foot every 1,200 years. That works out to a yearly rise about the width of a human hair. Of course, there was no one around 20 million years ago to notice what was happening. But even if the place had been packed with millions of people all holding binoculars or microscopes or whatever, no one would have had the slightest inkling that the Earth was moving.

Mountains of rock collapsed into sand beneath the fury of the water and the wind and the liquid Earth. Millions of acres of stone moved, but no living being noticed. Invisible, untouchable, silent, tasteless, distinctly odorless, slower than molasses. Surely, there were occasions when anyone would have noticed that something was happening—earthquakes, dramatic floods, collapsing canyon walls. But whole human lifetimes elapsed between those events. A hint now and then, maybe, but as unimaginable amounts of rock disappeared, as the Earth ripped violently at itself, no one noticed.

VICTOR

"Here, man, I brought you one of my pictures." He hands me a tinfoil and matchstick portrait of Jesus. He's done a great job. Even if I didn't know the portrait was supposed to be of Jesus, I would have recognized him immediately. Which in itself is interesting, since no portraits of Jesus were painted until hundreds of years after his death—long after everyone who might have seen Christ was dead and decomposed.

The work is meticulous—each stick laid and glued very carefully. Victor has used different lengths of matchsticks and sometimes stacked the sticks two or three deep to create the face. The burned match heads are gone and only the clean light-colored wood remains. Mostly it is a full-face view of Christ with a surrounding aura of matchsticks.

"Turn it over," Victor says to me.

I turn the portrait over and see that there is a poem attached to the back. The words are typewritten:

I remember our heme washed in meenlight
And your leve like a dewny deve.

"That's one I made for my girlfriend," Victor tells me. "She said she really liked it."

Victor's earlier mention of moonlight is my only clue as to what is going on here. Moonlight becomes meenlight. Apparently, through some defect with the typewriter, every "o" has been transformed into an "e."

"Did your girlfriend ask you anything about this poem?" I ask him.

"No, man. Why would she?"

BLIND SPOTS

We missed the Grand Canyon because it happened too slowly. And that's a lot to miss. But we miss even more because it happens too quickly. A vision doesn't form in an instant. It takes time to see something. Because of that, we never really see the beginning of anything. We just think we do.

In truth, beginnings are beyond us.

And even as the story unfolds itself, much of what happens disappears—like the individual frames of a movie as the quickness of events fools us into seeing motion where there is none.

Surprisingly, that doesn't seem to bother most of us. Somehow, in spite of our blindnesses, we find our beginnings and we fill in our gaps—sometimes with no regard for the facts.

In all right-handed people and most left-handed people, the ability to speak or write language involves primarily the left hemisphere of the brain. Normally, a bundle of nerves called the corpus callosum connects the right and left hemispheres of the brain, so visions seen by either the right or left hemisphere transfer to the other side and we can easily speak of what we saw. But there is a way to prevent that, to eliminate the facts from our words.

One approach to the treatment of epilepsy involves severing the corpus callosum—splitting the two halves of the brain off from one another. Because of that, several people walking around today have noncommunicative right and left hemispheres. Surprisingly, after their surgeries, these people seem no different from you or me. But an interesting experiment done with these people suggests that facts—facts about beginnings, facts about endings, all facts— sometimes play a very small role in stories we tell about what we have seen or done.

For most of us, if we are shown a word on a computer screen—using either our right or our left eye—and then asked to identify the image it describes (for example, the word "bell" and then select a picture of a bell from several other images), we can do so quickly and accurately.

But when people with severed corpus collosa are shown a word on a computer screen in a similar way, things change. If the image is seen by the left side of her brain, a woman with a severed corpus callosum can speak of it normally and explain her choice and its relationship to the word on the screen. But when the image is seen only by the right hemisphere, something very different happens. If only the right hemisphere sees the picture, the same woman can point to a similar object with her left hand, but she cannot explain in words what her left her hand is doing or why.

Interestingly, faced with the inexplicable behavior of their hands, these men and women do not resort to defeat and offer "I don't know why I did it" sorts of statements. Instead these people very quickly construct stories to explain what has just happened. The stories, of course, have no relationship to the images the people have seen, no relationship whatsoever to reality. The stories are confabulations or, more directly, lies—lies told to patch holes in the universe. But for many, perhaps all, of us that lie is preferable to the hole—better than having no explanation for what we have just done.

And it isn't only surgery that does this to people.

Herpes virus encephalitis—an infection of the brain by the same virus that causes cold sores—can sometimes destroy the hippocampus. The hippocampus is a part of the human brain that plays a major role in memory. People without functional hippocampuses seem very normal on the surface. But if you ask these people to remember something for more than a moment or two, they cannot. Interestingly, these people seem to retain the ability to understand the meanings of many words and related categories of things. For example, a person with a damaged hippocampus might well understand what the word "automobile" means, but he cannot recall what sort of automobile carried him here this morning. But if you ask such a person to tell you about the car that brought him here, he doesn't hesitate. Instead, he makes up a story based on some sort of car he might have once ridden in.

When faced with our own ignorance, our own inexplicable behavior, all we have is our stories.

VICTOR

November.

I confront Victor over his efforts on reading and writing.

"Sure, I want to learn to read and write," he answers without a moment's hesitation.

"You're sure?"

"Absolutely sure, man."

"Really?"

This time he pauses and looks at me with a little smile on his brown face and says: "What would you do if I said no?"

"Well, I guess I'd quit wasting your time and mine," I answer.

"What if I didn't think this was a waste of my time?"

"Huh?" I say.

We're sitting in the weight room downstairs now. It is too cold to meet outside. Victor straddles one plastic-covered bench, I'm atop another. Barbells, big enough for Kong, sprout like mushrooms from the dark spaces in the room's corners. There's a treadmill, one more path to nowhere, against the wall, and a couple of benches for some kind of bodybuilding. Talc covers the grips on the long iron bars between the weights, and the room reeks of human sweat. The light here is too little even for a space this small.

"I get out of here six months earlier if I try to learn to read and write."

"Regardless of whether you learn to read and write, right?"

"That's it, man."

"How soon do you get out?" I ask.

"Three months from now, if you stick around."

He looks at me. I look at him. I give up one hour a week for three months. Let's see, that's roughly twelve hours total. Victor gets six months of freedom.

"So what do we do?" I ask him, laying aside the books.

"Let's talk," he says and smiles his tight smile, a curious look in his eyes.

"About what?" I ask.

"I could tell you about prison."

There are few fears I have that will pinch my arteries like my fear of prison. And there are few places on the face of this green Earth that I know less about than prisons. I'm here for the duration, and Victor knows it.

December.

Now when we get together, Victor and I carry our books to the concrete room in the basement and Victor tells me stories.

"I never had a chance, man. The lawyer they gave me was no good. He pled me to eight years before I even had time to think about it. I served five and a half of those years before I came here, man. That is a long, long time. "Cañon City is a hard place, bro."

Victor is referring to the Cañon City Correctional Complex in Cañon City, Colorado. The complex is home to nine state and four federal prisons including the "Supermax," also known as the "Alcatraz of the Rockies." In Fremont County, Colorado (where the correctional complex is located), 16 percent of the total population is in prison. Most of the rest the rest work there. The Supermax is home to the aforementioned Ted Kaczynski, Timothy McVeigh (one of the Oklahoma City bombers), Ramzi Yousef (the World Trade Center bomber), and Zacharias Moussaoui (the only man convicted for crimes associated with the 9/11 World Trade Center attacks). I am deeply impressed that Victor was in with that crowd.

"What's it like inside?" I ask, like a kid in some second-rate James Cagney film.

"It's bad, man. It's really bad. The gangs run pretty much everything—black gangs, Mexican gangs, even Indian gangs. They look out for one another. If you ain't in a gang, you gonna get banged up some. But the baby rapers, they get it the worst, man. Everybody hated them for what they'd done. They couldn't get into no gang, so everybody took it out on them."

I am afraid to ask just what that means, not the "baby raper" part, I think I understand that. It's the "took it out on" part that I don't know about, maybe don't want to know about.

"Yeah, they got it bad, man," Victor says as he strokes his lower lip.

Instead of pushing the baby-raper button right now, I voyeuristically ask him how he, a drug dealer, was treated. He's wearing a sleeveless white T-shirt and faded blue jeans. He swipes his hand across the thinning hair at the front of his head. It takes him a breath or two before he seems to hear what I asked. "First day I was in, I ask around about who was the toughest guy in my cell block. Next day, I went and found him and before he could say anything, I hit him in the mouth as hard as I could. Then he beat me up real bad, and they threw me into solitary for six weeks."

Victor is only about five feet seven inches tall, maybe 180 pounds, and though he looks stocky and like he can handle himself, the picture that forms in my head is not pretty.

"The day I got out of solitary, while the guards were taking me back to my cell, I passed the same guy in the hall. I hit him in the mouth again, hard as I could. He didn't go down, but this time the guards stopped him before he could do much to me. He was plenty pissed, though. They threw me back into solitary, but this time, I wasn't all beat up. So it wasn't so bad. They gave me twelve weeks, but I'm looking at eight years, so I figure I got to do something. It gets pretty weird in solitary. By the time I got out, I was ready to do it again. But the guy offered me a truce. After that, most of the others left me alone, too. Things got better after that.

"I did okay. Man, though, the baby rapers, they had it bad."

Size Matters

Now, for me, the invisible has become an obsession. I lie awake nights thinking about what I am missing.

Someone who had some extra time on his or her hands once figured out that that there are about 10^{29} bacteria in this world. A number too big to imagine. But that number means that everything in this world, including us, is crawling with bacteria. We're rotten with them.

From a mile or two above the surface of the Earth to a mile or two below it, layers of bacteria. From boiling deep-sea vents to arctic ice, layers of bacteria, crusts of bacteria.

A single bacterium weighs about 500 femtograms. A femtogram is 10^{-12} grams. A gram is 1/28 of an ounce. Unimaginably small, irretrievably invisible. A myriad of tiny nothingnesses. But when you put them all together, the result is staggering.

A fully loaded aircraft carrier weighs about 100,000 tons. It would take a billion fully loaded aircraft carriers to equal the weight of the world's bacteria. The total area of the flight deck of an aircraft carrier is 4.5 acres. So, 1 billion aircraft carriers would take up 4.5 billion acres. That is enough aircraft carriers to cover the United States three deep in aircraft carriers—aircraft carriers from Portland, Maine, to San Diego, California, to a height of 732 feet (about 72 stories)—and all of it invisible.

And what about things that are just too big for human eyes, the far end of the road or the river, the county, the state, the country, the continent, the Earth, the ocean, the moon, the wind, the solar system, the Milky Way Galaxy, this universe, and others?

And then there are all the other things we cannot see that we don't even know we cannot see because we haven't discovered them yet.

The scope of our blindness is astronomical.

VICTOR

"We used to run prostitutes right here in Fort Collins. Pick 'em up downtown and drive 'em out to the motels near the freeway."

"Wow, no kidding? Here?" I ask, agape.

"Yeah. Friday and Saturday nights. Lots of business."

"You ever get caught?"

"Me? What was I doing wrong? I was just giving some girl a ride, man."

"Did you make much money at it?" I want to know.

"Not enough, man, or I wouldn't have been driving in from Cheyenne with all that coke."

Throughout these conversations, we sit with open books in our laps, pencils at the ready. And periodically, someone comes to the glass windows that line the front of the room and checks on us. Satisfied that we are deep into our verbs, they move on.

I have thought about trying to write down some of what Victor tells me. But I am not sure what Victor would do if he realized I was taking notes about what he was saying here. So I don't. As much of it as I can, I load into my head and write it down later at home.

"So what are you going to do when you get out of here?" I ask him one January evening, the two us once again straddling the weight benches, the room smelling like moldy towels.

"I don't know yet," he says. "But I know one thing. I ain't ever going back to prison. Man, I hated that place. The things that happened to people in there, man, no one would believe. The stuff they done to us. When I get out of here, I ain't ever coming back, not ever."

Victor's upper lip rolls and moistens. He reaches with his tongue to wet his front teeth. I change the subject. "What do you do during the day around here?" I ask.

Victor's wearing cowboy boots today. I've never seen him in anything but sneakers. These boots look like ostrich skin, yellow with black uppers.

After a second or two, he says, "I got no choice. I work every day. Can't take no sick days. Can't take no vacation. If I'm not working, I'm back in Cañon City. Just like that." Victor shakes his head and watches his feet move inside of their boots.

Victor stops for a breath.

"So what are you going to do when you get out of here?"

"I don't know yet, man. But the guys I was working for when I got busted, they told me just to come and look them up. They were real pleased that I never named nobody else in this thing. They told me that. Told me to look them up whenever I got out."

"The guys you were moving the drugs for?"

"Yeah, the guys I worked for."

Victor is due to get out in just a little over a month now. I imagine he might have a chunk of his old drug money stashed somewhere, maybe even a lot. Or maybe, after he kept their names out of the whole thing, these men really are happy enough with Victor to set him up after he gets out. One way or the other, Victor clearly doesn't seem to worry too much about how things might be when he gets out. He just wants out.

"I thought you said you were never going back to prison," I remind him.

"I ain't. I don't plan to do nothing illegal. But I figure maybe they can still help me. 'Just look us up when you get out,' they said."

"When did they tell you that?"

"Oh, they came to see me pretty quick after I was in Cañon City. Wanted to talk with me about how things were going."

Refrigerator Blindness

Sunday morning, and I am searching in the refrigerator for a bottle of jam for my toast. I look over each of the white shelves—salsa, mayonnaise, butter, eggs, last night's lettuce, but no jam.

"What are you looking for?" my wife Gina asks.

"Jam," I say. "But I guess we don't have any."

She steps behind me, reaches in, and lifts a bottle of apricot jam from the shelf immediately in front of me.

There are two possible explanations for what just happened. The first I like to call the Callahan Theory of Temporal Displacement, or CTTD. According to this theory, things sometimes slip a few moments, possibly an hour or even a day, into the future. So when you look for them in the present they aren't there. But if you wait a while and look again, those things will have reappeared in this (now) future time in the same place where you just looked for them. I have experienced this with pens, pencils, keys, and wallets. So I like to think that the jam truly wasn't there when I was looking for it.

The more mundane explanation is that the jam was there all the time, but I just didn't see it.

If that's true, and my wife swears it is, then that narrows my visible world even further. Beyond the limits of visible light, problems with size, and the molasses-like movement of the millennia, I also have a personal set of blinders that further narrows my window on the world. If I don't expect it to be there, if I don't expect it to be that color, shape, or height, if I haven't learned how to see it, or I simply don't care to see it, it disappears. Another set of holes.

Not Quite the Last of Victor

In April, Victor and I have our last session. He tells me he wants to get serious about reading after he gets out, and he asks for my phone number. I give it to him and I walk out of his life. I think about Victor a lot for the first few months, even imagine he might reward me with some of that drug money for helping him go free. But months pass, then a year, and I hear nothing from Victor.

A couple of years later, I again volunteer for an adult literacy program. I end up working with a group of students who hope to pass their GED equivalency tests and earn high school diplomas. This is not at all like it was with Victor.

For two years, my wife and I work for this program, and then things just seem to peter out. Classes end one summer, and we don't get asked back in the fall.

Later, I find out that that is because they didn't want to take advantage of us, overuse us until we burned out. I don't feel burned out. By the time I find out why they didn't ask us back, I'm involved with other things. So my brief career in adult literacy dies a natural death.

One Sunday morning, a couple of years after I have left the illiterate to their own devices, I pick up the newspaper and pull out the section with the national news inside. As I do, the second or "local" section of the paper drops onto the kitchen floor. On the cover is an article titled "Fort Collins's Ten Most Wanted Criminals." Curious, I pick up the paper and begin to scan the article. Fort Collins's finest hope that publishing these pictures will encourage our upstanding citizenry to assist the police with their work. I just want to see what the other half has been up to. But before I can read much of the article, I notice that below the story, they have printed pictures of the ten most wanted along with a list of their crimes.

Number three is Victor Delgado.

Between his first and last name, in quotation marks, is the word "Skeeters." He is my student, and he is wanted for jumping bail, of all things, as a repeat offender for child molestation—baby raping. Number three most wanted. Right now, on the street, molesting children, and I put him there six months early.

The next day at work I turn on my computer and go to the Fort Collins City Police website to see what I can find. I locate the ten most wanted list and the accompanying pictures. Victor's picture already has the words "in custody" stenciled across his face. Nothing else. A week later, I return to the city website, and Victor is gone altogether.

Almost.

A few years later, I go the National Sex Offenders Registry and search for Victor. Sure enough, his picture and his address now adorn a page of the registry. But the record incudes nothing about his convictions. That draws me to the Colorado Court Database. There, Victor's name appears twenty-five times. Mostly driving-related stuff—driving without a license, DUI, driving without insurance, and the like. But in July of 2001, the people of Colorado charged Victor with sexual assault by a person in a position of trust on a child under fifteen years of age. For that, Victor was sentenced to another four years in prison.

I never do find out just what put Victor in prison the first time. Maybe, in spite of what he said, it didn't begin in Colorado. Who knows? Regardless, that's where the story ends, mostly empty.

But the story is all I have.

_____

[Because of the nature of this story and the implied confidentiality of conversations, I chose to use a fictitious name in this essay]

THIS IS NOT THE END

Facing Up to Our Immortality

Every day for billions of years, this world has tested every gene we carry. When genes failed those tests, people (or creatures that might have one day been humans) died. That makes for very powerful and very useful genes—to a point. And that point comes when we are no longer able to reproduce. When we can offer no more sperm or eggs, we have climbed out of the gene pool. No longer Darwin's children, we must now fend for our selves, fight for our "I"s in a place we know nearly nothing about.

The first three-plus billion years of evolution left our ancestors with nothing to offer beyond reproductive years. Well, almost nothing. Even among the first proto-humans, the old could help to protect and feed the young. That increased the likelihood that genes passed to the young would one day find their ways into more and more people. Among primate species, that is no small matter, since most of us are born defenseless, unable to feed ourselves, and take years to achieve self-sufficiency. But protecting our young gave us no significant advantage over many other species. And remember, evolution is all about advantages. A few tens of thousands of years ago, that changed.

Human language unlocked a Lamarckian door that opened into a place where Darwin himself had lived but never noticed. In that world, our world, what we learned could live beyond us. For the first time old "I"s had a path into the "I"s of children. Once we spoke, we took a first step toward immortality and a non-Darwinian way of evolving. Darwin quickly caught up with us. The survival advantage of sharing what old "I"s had seen and learned was huge. There had never been anything quite like language, and those who spoke it spread. Along with that spread there was abruptly a selective advantage to longer-lived humans. Beyond reproduction, we could still pass on our stories to our children, to our grandchildren, and beyond—genes shifted.

Then we wrote and cheated death itself. Oral traditions lack fidelity. So do written traditions, but less so. Our thoughts became immortal. Lamarck had foreseen that. Darwin missed it.

But our bodies remain mortal. Aging and death still hold us in their calloused fingers. That's important. Aged "I"s do things young "I"s would never dream of—sometimes bad things. And the fading light of life can change the images old "I"s see. Change the stories that carry those "I"s.

Facing death, Robert Falcon Scott chose to carry forty pounds of useless rock across hundreds of miles of ice. The rocks were still there when rescuers dug Scott's frozen body and his sledge from the Antarctic ice. My father chose things much heavier and much harder than rocks to burden himself with on the trek to his deathbed. Near our endings, "I"s may shatter under their own weight. But that is about to change.

For nearly as long as there have been people, we have honored and feared death—the ultimate and inescapable consequence of living. Priests and poets feed on it, so do worms and bacteria. Only two things in life, so the saying goes, are for certain: death and taxes. Maybe not. Maybe only taxes are inevitable.

For eons, immortal beings have walked among us. We just didn't notice. Science has remedied that. Nearly as soon as we can hear, we learn that life is short and that all living things must die. That is simply the way things are. Maybe not.

The death of "I" seems built into us not because of some immutable rule of life, but because of a few genes evolved to help prevent primitive human beings from clogging up their gene pool and eating themselves out of house and home. The mortal insistences of those genes are written in their codes. But codes can be broken and genes tailored to look more favorably on immortality. The finality of "I" may be just a matter of circumstances created by a few old genes, ones we no longer need. The future of "I" is looking up.

Lamarck's giraffe

9

The Mysterious Visions of Jean-Baptiste Pierre Antoine de Monet, Chevalier de Lamarck

We often give Darwin most of the credit for explaining
how we humans came and continue to be. But Lamarck's
giraffes (frequently discounted as the musings of a confused
fool) may have more to tell us about parts of human lives
than Darwin's finches.

We're seated in circled desks, so that each of us can see everyone else.
We are here to discuss where selves come from. The seating arrange-
ment helps to lubricate our discussions. At times, it lubricates other
things as well. Janine, a dark-haired, attractive woman, has worn a
dress to class today. Just now, though, she seems to have forgotten
that. Though I cannot see her from where I sit, Chad, another twenty-
something student—seated directly across from Janine—has noticed.
From now until the moment Janine crosses her legs, Chad will hear
nothing I say.

Most of us seated in this room (and probably seated or standing any-
place else) imagine that we sit here as individuals with a history that

DOI: 10.5876/9781607322337.c09

spans at most a few decades. But it isn't that simple. Each of us walked into this room dragging a four-billion-year-long tail cleverly disguised as a set of chromosomes. Written in those chromosomes are words of lust and hunger, murder and revenge, fear of the dark, and an abiding love of life. Each us, too, carried into this room a 100,000 (or so)-year-long tale of what is means to be a human being—the stories of our ancestors cleverly disguised as seemingly personal thoughts.

That's the part most of us forget—the stories passed down over eons from one person to another given to us at birth as surely and as inevitably as DNA. And though forgotten, these stories are as essential to our selves as those chromosomes.

I'm a biologist. I believe human behavior reflects human history, especially the history gathered up inside of our genes. That's how I learned it. I am the product of DNA, inescapably.

We are here at this moment because every one of our ancestors—fish and beyond—ate, didn't get eaten, and followed an irresistible urge to find a mate and reproduce.

We have no problem seeing that clearly in other species. For example, when a male chimpanzee mates with a female in estrus, we call that instinct. There is no supposition of soul searching or rational thought by the chimp. It simply does what it does because, if its ancestors had not done the same thing, this particular male chimp would not be here right now. Instinct driven by genes tested over billions of years for their utility.

We don't imagine much, if any, human behavior is instinctual. Even though we share something like 98 percent of our DNA with chimps, we're simply not like that. Ours are reasoned decisions and carefully considered actions.

In reality, not so much.

Most of the things we do, we do because we once had to do them to survive. Well, survive and reproduce. That's what the last four billion years of evolution has been all about—survival and reproduction. And that past, that genetic past, consciously or subconsciously, governs every move we make.

That's what Charles Darwin taught me and right now, Chad is verifying that. But consciously or subconsciously, there is another force at work as well, a force that is so apparent we almost always ignore it—a force full of human words.

Perhaps it's best to begin at the beginning.

THE BIG BANG

Actually, this story begins just before the Big Bang. There may be stories that begin even earlier, but I don't know any of those stories. Just before the Big Bang, every point that is me today overlapped completely with every point that is you today. My father and I were one. My mother and I were one. My wife and I were one. Even Orrin Hatch and I were one.

We lay, the bunch of us, far heavier than lead in a single submicroscopic bed. We had no differences of opinion or even of locale. Imagine that, me and Orrin in complete agreement. We wished for only a single thing. We saw all things with a single eye. We were of one mind, as close as we ever would be.

One day, all of that changed. With no warning, space-time erupted into itself. Suddenly, there was no one. Just the screaming plasma. None of us overlapped with any other now, not even with ourselves. Hurtling toward the edge of infinity, we danced.

Eight billion years slipped by.

The darkness broke, and flames began to burn here and there in the emptiness. The universe split—part fire, part void. In the hearts of those fires, thermonuclear fusion slammed hydrogen atoms together and helium appeared. As the first smaller stars died, helium was hammered into oxygen and carbon. Later, as more massive stars died, within their corpses pressures reached indescribable levels, and the bigger elements—copper and zinc and gold—appeared. Elements we would need soon.

Once things cooled a bit, we gathered up the stuff of life in our knapsacks and met in a far corner of the universe. We were a still, dark cloud at the edge of a young galaxy. Uncertain of ourselves, we waited.

Then, slowly, and nearly against our wills, I, my mother, my father, my wife, Joe DiMaggio, and Stephen Hawking—all who ever have or ever will walk this planet—coalesced into a small, rocky orb, 93 million miles from the nearest star.

It was hot. The radioactive elements that we brought with us were falling apart so fast that even rock could not withstand the heat. The surface of our world was liquid stone. Water was vapor roiling in black clouds above a world without a sea, without a mountain, without a blemish.

One hundred million years, maybe more, passed, and the radioactive elements decayed into stable, cooler isotopes. A thin rock crust formed across the surface of the molten Earth, and through that crust, volcanoes pushed up the continents. Asteroids brought water to the cooling planet.

It grew colder still. Above the rocks and the land, the water vapor condensed and it rained. For a very long time, it rained.

We washed into the seas.

There, the gift of life began to unwrap itself.

We became molecules, chains of atoms in a warm sea. We did what we could.

Mostly that involved reproducing. We just reached out and gathered what we needed from the primordial soup and—like snapping together LEGOs—made more just like us, well, almost just like us. Mistakes were made.

After a while, the raw stuff we needed to make copies of ourselves was nearly all used up. First uracil, the one base building block unique to RNA, ran short. Then the others disappeared, adenine, guanine and finally cytosine. Life slowed. Until one of us found a way to dismantle other molecules—a sort of molecular cannibalism. From those pieces came more of us with the ability to dismantle.

The best eaters and best reproducers made more of themselves, quick as fire chews through dry grass. Until, once again, the seas were full of us. Again, the pace of change lessened. Then, one of us figured out how to wrap herself in a bit of oil and protect herself from being dismantled by other molecules. The first cell was born.

Time and the sea delivered us up: whole, hungry, and lustful.

From that simple chemistry and those basic needs would come me, you, and all the rest.

We became bacteria, then cyanobacteria. We took sunlight and used it like a blade to slice carbon from carbon dioxide, pound it into sugars. We belched up oxygen—the stuff that would one day rot iron and ruin flesh. This new oxygen immediately oxidized all the iron in the oceans and filled the skies. The brown seas turned azure and the pink skies turned blue.

We ate starlight.

Abruptly, we were everywhere. We fed. We excreted. We anabolized. We catabolized. We were archaea and bacteria—jacks of all trades, masters of none. There was no place, yet, let alone time, for thought, sight, sound, touch, taste, motion, breath, or poetry. But we had mastered survival. We flooded the seas and fed on one another.

Communes formed—communities of bacteria learned to live together for the common good, one inside the other. Eukaryotic life arose—cells with internal membranes. A room for DNA, a place for fire, one place to build things, another to tear them down. Whole genomes were shared and mixed. Some blue-green algae became chloroplasts—the parts of green plants that turn sunlight into molecular energy. Other bacteria became mitochondria—the parts of our cells that burn sugar and oxygen to make our lives possible. Life became symbiotic.

Time unwound itself. We were jellyfish, translucent specters of what was to come. Multicellular animals with the potential of gods. Now there was a path and a place for eyes, and fingers, nerves, toes, livers, lungs, blood, and bone.

We fought with one another almost constantly. We fought for food, we fought for mates, we fought to live. Sometimes we won, sometimes we lost. We lived, we died.

Abruptly, we were fish, vertebrate, cartilaginous, learning new ways to kill. The oceans filled with us. Fish of all kinds—fish with bones, fish with jaws, fish without jaws. Fish lit with lights and fish covered with scales.

Then we were salamanders. Spotted, slippery, stub legged, and thick witted. No longer tied to the sea, we moved onto land, breathed oxygen unmuddied by sea water. We grew lungs. That was novel, at least at first—breathing air, walking around on legs. But after a few million years, it wasn't enough.

We became lizards. Great horned and striped creatures like tetraceratops and dimetrodon, armored with scales and razored with claws. We ate, we rutted, we lolled beneath the ferns while others' bones rotted in our bellies.

Then we took a gamble, one we nearly lost. We became mammals—little shrews, at first. Warm-blooded, furred, milky, and misguided. We called ourselves megazostrodon—large animal tooth. The name made us feel better about ourselves.

Great lizards ruled the world then. Huge carnivores and even larger herbivores careened like broken buses through the valleys and swamps. But we were too small for the behemoths to bother with. We hardly made a toothful for beasts so great. We darted from hole to hole beneath their horny claws.

One day, the sea boiled over, and there came a great wind. Water, then fires, raced across nearly all of our world, leveling the forests, leaving only ash and stubble. Then darkness, a darkness untouched by sun or starlight. We huddled in our burrows, eating ice and taking what small heat we could from one another. But we never loved the darkness, or one another, for that matter.

A meteor, a mere 6.2 miles across, had slammed into the sea near what is now the Yucatán, and the face of the world was changing. The fire, the darkness, the cold were nearly unbearable. But that fire and that ice forged men from mice. In the cold, the great lizards faltered and began to die. But our warm blood carried us into a new age. Now was our time to be something other than shrews and mice, something other than fodder for half the living world. Our gamble had paid off.

We moved into the trees. It was safer there. Fewer things fed upon us in the dark. We became primates—monkeys and apes. Large, strong enough to fight for what was ours. Strong enough to kill. We lifted stones, we gathered together.

We moved back onto the ground and we became men and women, but not *Homo sapiens*, not yet. But we were *Homo*, *Homo habilis*, and overhead the African sky glistened with starlight. Now the only sea we knew we carried within us. We never spoke. Our thoughts were pictures of the world—things we might eat, things that might eat us. But the colors of the pictures were changing. We had axes now—sharpened stones heavy as heads. Our jaws grew smaller and our brains grew larger. We worked together and the world spun on its axis.

Nothing was beyond us. We had overcome every challenge. We were rich and strong and fecund.

We became *Homo sapiens,* Neanderthals. Sapient.

We spoke, and the Earth shuddered.

We made music on bone flutes. We carved bits of rock into the shapes of women. In Chauvet, by firelight I painted a bear. It was full of me, strong, urgent. You sketched a rhinoceros. I admired it.

We raised goats in Iran and pigs in Thailand. We grew wheat and barley in Iraq. And in Egypt, we counted on our fingers the number of times the sun rose and fell. We sang songs to our children, songs full of ourselves and of the past.

We rolled papyrus into sheets and our thoughts fell onto pages and our words reached beyond us in ways never imagined.

We made alphabets in Palestine and Syria. We grew corn in the Americas and smelted iron in Britain. We built the Great Wall.

Rome rose and fell.

Then we nearly died when the plague raged across Europe. Another test, always another test. Once again, all of our efforts nearly came to nothing. We carved the Pietà.

Five hundred years later, in muddy fields and in our cellars our potatoes rotted. We left Ireland then, for another world. New York City chewed us up and spit our remains onto steel rails. We followed the rails to Kansas.

In Kansas, some of us died. I was born.

THE PURPOSE OF THE PAST

From such a simple beginning came 30,000 or so of the finest genes that four billion years and geochemistry could cast. The human genome.

Every one of our direct ancestors survived, at least long enough to reproduce. Every worm, every jellyfish, every lizard survived. We know that simply because we are standing here. Someday reproductive technology may reach beyond this. But for now, our existence is proof of our ancestors' skills. And that is their gift to us, survival wrapped up inside twenty-three pairs of chromosomes.

For millennia, evolution has been shaping human genes for a single purpose—reproduction. Those who didn't measure up didn't reproduce. Evolution is a harsh mistress. Genes that helped animals make more of themselves survived. Genes that didn't help quickly sunk to the bottom of the gene pool and were eaten by scavengers. Survival of the fittest. Over the years we grew smarter. We grew stronger. We grew warier.

At the moment of conception each of us is given that legacy—helices full of genes old as life itself. Genes that brought fish from the dark into the light, genes that made lizards strong, genes that allowed apes to stand up, genes that crushed others and forged a living from their remains, genes that will write poetry and explore constellations, genes that will stare at the stars and wonder, and genes to make us care about all of it. A genomeful of genes.

All of it, more or less, as Darwin described in *The Origin of Species:*

> It is interesting to contemplate an entangled bank, clothed with many plants of many kinds, with birds singing on the bushes, with various insects flitting about, and with worms crawling through the damp earth, and to reflect that these elaborately constructed forms, so different from each other, and dependent on each other in so complex a manner, have all been produced by laws acting around us. These laws, taken in the largest sense, being Growth with Reproduction; inheritance which is almost implied by reproduction; Variability from the indirect and direct action of the external conditions of life, and from use and disuse; a Ratio of Increase so high as to lead to a Struggle for Life, and as a consequence to Natural Selection, entailing Divergence of Character and the Extinction of less-improved forms. Thus, from the war of nature, from famine and death, the most exalted object which we are capable of conceiving, namely, the production of the higher animals, directly follows. There is grandeur in this view of life, with its several powers, having been originally breathed into a few forms or into one; and that, whilst this planet has gone cycling on according to the fixed law of gravity, from so simple a beginning endless forms most beautiful and most wonderful have been, and are being evolved.

More or less. In the very act of writing this Darwin overlooked the force of words and the genius of Lamarck.

At the moment of birth, along with forty-six chromosomes, we are also given another piece of humanity, a piece smelted in the same fire as our genes—human stories.

THE CIRCUMVENTION OF CHARLES DARWIN

Those stories shape us as surely as our genes. More important, those stories will carry us beyond our genes.

Once we can no longer bear or father children—our genes are ours to keep. We will pass them on to no one. Now we are on our own. No gene that saves us, no gene that kills us, will ever change another's DNA. Now there is only unnatural selection.

Abandoned by Darwin, we live in a true terra incognita. A place our genes know nothing of, a realm completely unexplored by and of little interest to evolution. In that place, we are at the mercy of a biology freed from its own history, a biology with no fear of the future, biology in its rawest form. A place few of our ancestors ever found themselves.

You might argue that life beyond reproductive age still offers a certain kind of protection for some of our own genes running around inside our children or grandchildren. And maybe that is some of it. But what is so special about 70 or 80 years of living? Why not 60 or 120 years?

It isn't just about genes. There is more to it than that.

THE POWER OF MYTH

Jean-Baptiste Pierre Antoine de Monet, Chevalier de Lamarck was born August 1, 1744, in Bazentin-le-Petit, France. He was the last of eleven children. Perhaps his struggle to find a place so late in a family so large fired his interest in evolution and adaptation. Regardless, after a period in a Jesuit seminary, a stint in the military, and a brief career as a bank clerk, Lamarck fell in love with botany and medicine and the diversity of living things.

Around 1801, he began to publish the ideas that would make him so famous or, more accurately, infamous. Lamarck had seen the ways that animals and plants changed with time. He was convinced of the reality of evolution. And, rather ingeniously, he proposed that the stressors of everyday life drove evolution. Change, he argued, occurred when an animal or a plant was forced to struggle for survival, to reach beyond its comfort level to live and reproduce. Then, somehow the changes that came from the struggle to survive were passed on to sons and daughters, or calves, or foals, or kittens—evolution through inheritance of acquired characteristics.

The classic example was the giraffe. Giraffes, Lamarck proposed, had long necks, because their ancestors had to stretch their necks very hard to reach the higher leaves on trees—the only leaves that remained after all the shorter giraffes had fed. The constant stretching made the giraffes' necks grow longer and the long-necked giraffes then passed that trait on to their progeny.

Today Lamarck's work is most often presented as sheer foolishness, as an example of how wrong someone can be about how evolution works. The bumps, bruises, and stretches of a lifetime do not change genes, particularly the genes present in sperm and egg—the only genes of ours that ever reach our children. Giraffes' necks didn't get longer because they were stretching them all the time. Giraffes' necks got longer because one day, on some long-forgotten veldt in central Africa, a mutant giraffe was born. This giraffe's neck was much longer that his brother's or sister's or mother's or father's. This giraffe thrived, because he had all the food he could possibly eat—the high leaves that none of the other giraffes could reach. So the mutant giraffe reproduced early and often. And all of his sons and daughters had long necks. And they reproduced early and often. And so on, until all giraffes had long necks.

Lamarck had good intentions. He just wasn't as smart as Charles Darwin. Lamarck, though he tried, was just wrong. That's what our textbooks and our teachers tell us. And maybe Lamarck's genius wasn't meant for the Galapagos finches, but that doesn't mean he wasn't a genius.

At the moment of conception, our parents give to us their bags of genes. As Darwin imagined, those bits of DNA have a lot to do with who we become, but they probably don't change much in a single lifetime, not by stretching, not by running, not by wishing.

But at birth and even before, each of us is given another treasure—a library full of ancestral tales. As soon as we have ears, they are filled with our parents' stories. We get them first in lullabies and books read aloud, even before we are born. Later we get them from grandmothers and grandfathers and parents and we get them from the radio and the television, the toys that fill our cribs, and the mobiles that twirl over our heads, the words our parents share with us and the words they share with one another. And as we grow, so do our stories.

And like our parents' genes, these stories change us, literally, physically change us. As we listen to the words, our brains change shape and size. Then the wires that stitch together our neurons untie old knots and twist new ones. Children read aloud to, even while they are still in the womb, quickly learn to love the stories they have heard and to shun others. And infants and children who never hear words and stories lose the ability to ever understand them.

Words change our brains, forever.

Just like genes, stories change the course and the character of our lives. Unlike genes, stories are told and retold, shaped and reshaped throughout our parents' lives. By the time the stories are handed over to us, most of them are no longer the stories our grandparents gave to our parents. By the time we inherit them, like Lamarck's giraffes, the stories have acquired longer necks and all the other lessons of a lifetime. And we acquire more—our grandparents' original stories, the stories of our culture, the stories written in our books, the stories of our people. The human bibliome.

Lamarck may have been wrong about giraffes, but he was right about stories. Living changes our stories, makes them more intricate, more personal—better, perhaps. And then we pass those stories to our children and our grandchildren. Long after we are capable of producing a child, we are fully capable of weaving a tale. And we are capable, as well, of passing stories on to our children and our grandchildren and anyone else willing to listen. After reproduction, we may be abandoned by Darwin, but Lamarck quickly reaches out for our aging hands.

Beyond the age of reproduction comes the age of story, a time for telling and retelling the tales that have driven our lives. How to catch fish. How to make love. Where the Earth came from. Where it is headed. Why birds fly and worms crawl. How to make a flute. How to build a home. How to hear mountains speak. Why we cry.

It is the time for words.

And it seems even natural selection figured the importance of that one out, since, given the chance, most of us will live well beyond our reproductive years—and that, I think, takes genes. Though 100,000 years seems a very short time for dramatic genetic change. Perhaps, at first, longer lives simply allowed for greater protection of our progeny, our genes, our children and theirs. But words turned this time into much more.

Compared to genes, words are very young. But that really makes no difference. Though they came late, words changed everything. Words erased solitude. Words carried us beyond genes. Words pulled us from the trees and pushed us toward the stars. Words drew us closer. Words created gods and destroyed demons. Words made us immortal.

Our genomes made us breathe, our bibliomes made us ask why.

Even in Darwin's most lucid dreams, it seems he never imagined the power of story or the force of words. All the while, Lamarck was dreaming the dreams and seeing the visions of his father and his mother, and their fathers and their mothers, and . . .

Most of us learned somewhere along the line that we could discount the predictions of Lamarck and came to believe that the words we gather during our lives are inconsequential, powerless against the hammer of biology.

They aren't, though.

Humans began speaking to one another somewhere around 100,000 years ago. The words spoken then are unknown. But we are tied to those words as directly as we are to our ancestors' genes. And in among those first spoken phrases, as surely as in among those first genes, were forces that would someday save us and others that could someday destroy us.

We know that mutant genes can do terrible things.

Take Lesch-Nyhan disease, for example. A single mutation, a one-letter change in a boy's genetic code, and that boy will mutilate himself—chew off his fingers beyond the knuckles, bite off his lips, chew his tongue, and bang his head against a wall until he is bloodied. Given the chance, he will beat his mother or father. Apparently these children feel pain and love their parents as strongly as any of us. They just can't control themselves. A tiny chemical difference and the most basic of human rules—don't eat yourself—evaporates.

Mutant stories can do equally terrible things. Take the Holocaust, for example.

Whatever it means to be me or to be you is as bound up in our stories as it is in our genes. What we say to ourselves, what we say to one another, changes everything.

Janine has finally crossed her legs. I have Chad's mind back for a moment. He asks me to repeat what I just said. I should be annoyed, but I'm a biologist.

Robert Falcon Scott group. Photograph taken with a string attached to the camera's shutter release: Edward Wilson, Robert Scott, and Lawrence Oates (standing), Henry Bowers and Taff Evans (sitting)

10

The Rock Collector

Human lives are stories of collections—words, coins,
stamps, lovers—collections of things we once felt we must have.
But as we age, most of us find time for fewer and fewer of these
things. The last of life is an exploration, an expedition in which
our very selves are at risk. Now the things we choose to keep must
be of great importance.

January 16, 1912, six days before my father was born, Robert Falcon Scott
and his team reached the southern pole of the planet. "Great God," he
wrote in his diary. "This is an awful place." It was twenty-one below zero.
The wind was blowing at forty miles per hour and howling in his ears like
the dead. The men had walked for months in the worst weather on Earth.
Their ponies had died weeks ago. A poor choice—those ponies. So the
men themselves had pulled all their supplies over hundreds of miles of ice.
And now the wind and the sleet lashed at them as though they deserved
it. Few men have experienced the full, flat fist of misery. Scott and his
men were living it, or so they thought. Then things got worse.

DOI: 10.5876/9781607322337.c10

"I'm going to have my tombstone moved away from your mother's. Down to the other end of that row," Dad says to me.

Curiously, during his last few years, most of what my father and I share, we share in Perkins restaurants. Neither of us intended for it to end like this. It just has.

As usual, we are sitting across from one another in a leatherette booth, and the course of our conversation is as familiar as the floor tiles here.

No one else wanted him, not really. My father could be a handful. Five foot seven or so, 135 pounds, his nose broken one too many times by surgeons hoping to save the rest of his face, deep brown eyes under a thinning layer of silvery hair. And hands brown spotted with age and roughed up by years of steel flanges and oily wrenches. Handsome enough. But he can be mean—thoughtlessly, and sometimes thoughtfully, mean.

We're a lot alike, me and this old man. He and Mom came to live in Fort Collins to be near me and my wife, Gina. Two years later, my mother died. Deep into her dementia, my mother's heart gave out. Bless her heart. Her death nearly killed my father, and that angered him. Not because it nearly killed him, but because it didn't completely kill him.

Poking at the ham and turkey in his chef's salad, he says, "I don't want to spend eternity next to her. Not after what she did to me. It was a shitty thing to do."

My father is ninety years old. I'm fifty-six. People say we look a lot alike.

Since my father came to live in this city, things have changed. At first we both had great hopes and intentions. The day he arrived we spoke almost boyishly about all we would do together. But now it feels like we are unraveling, like the last of this winter will never end.

His left eye squints at me from across the laminated table top. Four years ago, in a fierce snow and ice storm, a virus tried to steal that eye from my father. He and I wouldn't let that happen. But treatment involved sewing his eye shut. That night, one-eyed, he drove off with my mother into the teeth of a Utah blizzard. Against the odds, he made it home, but that eye was never the same. The squint is permanent.

Right now, though, my father's other eye stares at me with the same feral indignity. We slide deeper into our conversation, a conversation full of thorns and scorpions. His sausage-like fingers push his salad around the chipped, white plate while I watch.

I try to divert him, before he really gets going. "How's your salad?"

He doesn't fall for it. "I know just when she did it, too. You know how I know? It was all she could talk about while we were dating."

"You want some more salad dressing?"

"And that morning, the morning after Shorty fired her, that was the first time she could get away while I was at work. If I had come home for lunch that day, my life wouldn't be so lousy."

Once, we were good men, strong men.

As the five of them—Scott, Edgar ("Taff") Evans, Henry Bowers, Lawrence ("Soldier") Oates, and Edward ("Uncle Bill") Wilson—neared the pole, Bowers spotted the first cairn. More followed, then a tent, and a black flag. Dog tracks and ski marks littered the area. After hundreds of miles of fighting ice and cold and scurvy and dehydration, they had arrived too late. After wandering for months across a barren and broken land, they had lost.

Roald Amundsen and his Norwegian team had reached the pole before them. At most, only a few weeks before. Tacked to the discarded tent, Amundsen had left a note for Scott asking him to deliver a message to King Haakon VII, the reigning King of Norway. Insult upon injury.

Inside Amundsen's tent were fur mittens, some clothing, and a sextant. The note explained that Scott was welcome to any of it. Amundsen needed none of it now; he was almost home.

After nearly twelve years of planning and preparation, Scott and his men were two weeks too late. The chill grew deeper and the little moisture left to them froze in their noses. Though each had begun this adventure as a man, on his way to the bottom of the world, every man had become a beast. They had no choice. Only beasts could have faced that cold and those winds, the endless miles of ice, and the long hours in the frozen traces dragging the wooden sledge. For weeks now, they had not been men. They had been muscle and blood, clawing their way over the ice. Ice in their eyes, their mouths, under their nails. Ice in their minds. Only their hearts continued to clench and unclench in human rhythms. But in that moment, at the pole, the ice that had followed them so relentlessly reached into their hearts. The beasts faltered.

———————————

My mother's death, even though dementia had been carving away at her for years, hit us both hard. Her funeral service was tough. Dad and I argued over nearly every aspect of it, except that it would be held in a Catholic church. We even argued over the costs of the cremation casket, urn, and the slot for her ashes. But in the end, he bought a cardboard cremation casket and small yellow urn. Then he bought an extra stone slot next to my mother's, a slot that would one day hold his urn and ashes. He wanted to be sure no one else would lie in that spot.

My father had never lived alone until that moment. His mother, his roommates, my mother had always stood between him and the rest of the world. He told me that now he was cold a lot, especially at night.

He hated how much he missed her, he hated everything about what had happened to her, and he hated dealing with the trappings of a life he would just as soon be without.

He had failed her, and I had failed them both. With no other place to entomb his anger, he unleashed it on my dead mother. I pointed most of it at my father. We had been so full of resolve when all of this began, so certain that, together, we could make things work—for all of us.

But the night my parents arrived in Fort Collins, the fear came with them. We had organized for months: sold their home, rented an apartment, moved and arranged furniture, wired phones, changed banks, moved funds—we even bought flowers.

We had planned for everything. But we hadn't planned for nothing. Mostly what we found was nothing. I wanted to blame him for that. Why wouldn't he let my mother be?

Now, eighteen months after her death, my father is no longer sure that the ultimate placement of his ashes, so close to my mother's, was such a good idea after all. "Not after the way she treated me."

My father never read a book to me. We never fished or camped or floated, swam, or canoed on a pond, lake, or river together. For the first twenty years of my life, it seems like we hardly spoke. I think I frightened him. Me, the third of four, who at age three nearly died because of a mistake made with a pot full of boiling water. Not his mistake, by the way, but a mistake all the same. For twenty years, we had an understanding and we waited.

But once he began to think I would survive, maybe even outlive him, things changed between us, and we found ways to take care of one another. I'd find myself in the middle of Utah and want to see him. He'd pack my mother up and drive 200 or 300 miles so we might spend a day together. More than once, I (often with my brother) delivered him to a hospital or a clinic and sat with Mom while Dad got rid of a burned-out piece of himself or collected a couple of new ones. We didn't talk about the past or how things might have been, but we looked out for one another.

Or at least we used to.

"Right after we got married, her and her sisters, Gertrude and Glenna, all of them ran off to Tulsa to get screwed. Isn't it funny that all of them are dead now?"

I push my plate toward the center of the table and lean back into the rolled-and-pleated bench. It is like watching an old man being beaten with wooden sticks, watching the bruises bloom beneath those sticks and being able to do nothing about it. Watching the skin split and bleed and bleed and bleed. It is like watching my father die.

We've come a long way together, me and this old man. He was the first of the Callahans to graduate from college. That changed everything for his children. Somewhere along the way, I became a favorite child—a curious mixture of expectation and dis-ease. When I first left him, moved out to California, he showed up often as he could and helped make a hard place livable. I miss that, that sense of willingness. I miss too the extra muscle of him, muscle that could carry me if need be beyond the storm.

"Glenna was the first to go, and she burned up. Burned up completely, I heard. The roof of her house caught fire and she pulled down the ladder to climb into her attic. Trying to put out the fire, I guess. But when she pulled that ladder down, she buried herself in flaming boards. Burned up, completely," he says, with what seems to me like a certain grim satisfaction.

"Then Gertrude, she died while she was *screwing* some man. Then your mother—the Alzheimer's."

"I think I'd go for the screwing if it was me," I offer, trying to push this river from its course.

He pauses for a moment, looks at me, then chuckles. But the diversion is temporary. "You know what the last thing she said to me was?" He gestures at me with his fork.

In spite of their crushing defeat, Scott, Oates, Evans, and Wilson celebrated by smoking cigarettes saved for weeks for just this occasion. Amundsen's men had enjoyed cigars.

Then the second team to reach the pole set about making its own measurement of time and sun. At once, Scott and Bowers discovered that they had actually walked past the true pole. Just then, though, they were too tired to do anything about it. The next morning would be soon enough to walk back and make camp at the pole. The men pitched their tents, and that night each man slept inside an elk-hide bag full of demons.

Two days later, after hours of tedious and time-consuming measurements, they found what they believed to be 90° south. In fact, they were still nearly 1,500 yards from the true pole. Considering the primitive instruments they worked with and the conditions they faced, −20°F and skin-numbing winds, their measurements were remarkably accurate. Still, they were wrong, nearly a mile wrong. Wrong and late.

At their own pole position, a photograph—taken using a string attached to the shutter release on a camera bolted to a tripod—shows only men. No beasts among them, now. The faces are tired, wind polished, and frost bitten. The eyes are empty. Their equipment, pathetic. In Oates's diary entry for that day, he complained that his feet were now often cold and wet.

Ten months before her death, my father and mother celebrated their sixtieth wedding anniversary. Sixty years, and besides the two of them, just my wife and I shared the moment. Truly, by then, my mother was already gone.

Now in a last-ditch effort to bury his loss, my father has convinced himself that once, over sixty years ago, my mother was unfaithful to him. And his demons have offered him details for this tryst that make it nearly unbearable for him.

My father is now certain that my mother did this to him because, as he says to me, it was all she could talk about while they were dating—how much she'd like to slip off to Tulsa with her sisters and get laid by some traveling salesmen there.

The absurdity of it goes to the heart of his and my failures.

I have told him many times that I think he has imagined this tale. He is unfazed by my thoughts. Instead, he has added to his tale a story of another indiscretion, one she purportedly carried out during her last few months in a nursing home—an infidelity with another of the inmates there, both of them in their late eighties. More nothingness. More spaces where there should have been something.

And because of my father's failing memory, whenever we meet, he needs to tell me again about my mother and her lovers.

I am only thirty-four years younger than he. Not so much, really. He is willing to share everything he feels and fears. He has no sense for the space between something and nothing. I am so frightened by him that I want to forget all I can, now, this minute. Otherwise, one day I know I will do the same. Something and nothing, and never notice what I have undone—spilled the sweet crud of my life all over someone.

I watch his eyes and wait.

He stops talking as the waitress approaches our table. She reaches over and picks up my plate. "Need anything else, hon?" she asks me.

Help, I want to say to her, *help*. "No, nothing," I mutter aloud instead.

My father grins up at her and coyly says, "You can't have mine yet," while he hunches over his plate. The waitress smiles broadly at him and pats him warmly on the shoulder. She's maybe fortyish with a solid waitress look about her. Then she reacts like they all do to my father. "Don't you worry, darling," she says. "You take all the time you want."

On her way back to the kitchen she turns toward me. "He's *cute,*" she says.

For a while on their return, things went well for the Scott party. A favorable wind filled the sail on the sledge, and they made remarkably good time across the ice. The sun still shone all day, and the temperature rarely dropped below −20°F. But Evans's fingers began to blister. He had had problems with frostbite on his first trip onto the ice. Ever since, the frostbite ate into him more easily. Later, his nose began to peel and darken, then his fingernails fell out.

Along their way, Scott collected nearly forty pounds of rocks, "geological specimens," he called them. His need was great. He'd lost the race to the pole and it nearly killed him. He needed something, something to anchor his sanity, to save himself from the dark pit of failure.

He chose rocks.

And, as they fought their way across the rotten ice, he gathered them relentlessly. Then he piled them atop the already-heavy sledge. Pulling that sledge took its toll on all of them.

By now the food was gone, and many long days of intense hunger still lay between them and their first cache of horsemeat.

By late January, daytime temperatures dropped below −25°F. The snow, so slippery only yesterday, turned to sand. The sledge dug into the ice and refused to slide. The Antarctic winter was gaining on them, and the beasts who had pulled so long and so hard to bring them here had left them at the pole.

Oates's feet grew colder.

Believe it or not, I choose where we eat. I had taken Dad to other places, but when they didn't have what he wanted to eat, Dad would sulk. He didn't much care for surprises.

He didn't like very many kinds of food, either. One night, over dinner at our house, he confessed he had never eaten squash—acorn, butternut, summer, winter, none. Never a squash. Dad didn't much care for things he had never eaten before. It took us some time to talk him into trying the squash. He was reserved in his praise.

In addition, my father disliked all Italian food, all Mexican, Thai, Vietnamese, Indian, Ethiopian, Spanish, Continental, French, Russian, Czech, Cajun, Creole, Greek, Hungarian, any kind of fish, or anything that cost very much. I settled on Perkins.

The night Mother died, we abandoned her. I had left Dad at his house after lunch with plans to see him the next day to help him buy some tools. He called me up at 8:00 that evening and told me he thought I had better come to the nursing home. When I got there, Dad told me he thought my mother was already dead. I held my fingers to her neck for a long while trying to find something I knew wasn't there. Resolved, I turned to Dad and told him his wife was dead.

He said nothing at first. Then he told me she had just turned over and quit breathing. I told the aide I thought my mother had died. She came with her stethoscope and convinced herself that I was right. When the aide left, my father wanted to leave as well. I went with him. In spite of all of our plans, we left my mother's body to the whims of others.

On February 17, Evans, frost-bitten and exhausted, died. The following day, Scott and the other three reached a cache of frozen meat where they had slaughtered their ponies as they headed toward the pole. The horsemeat cheered them. After that, for a few days things went pretty well again. The temperatures warmed, the wind came from the right direction, and the team made good progress.

But the trip had taken its toll. And now, for the first time in months, the sun began to dip briefly below the horizon. Temperatures fell. Now Oates's toes were black and shiny as leeches. The flesh of his fingers and nose grew ragged from repeated freezings and thawings. At forty-one degrees below zero, the snow once again turned to grit, and the sledge froze. On March 10, Oates asked Wilson if he had any chance of surviving this ordeal. Wilson said he didn't know. In fact, Wilson knew full well that Oates would never make it. A week later, Oates asked that the men leave him behind, but no one would hear of it. The last day, they dragged Oates forward with them and then put him back into his hide sleeping bag for the night. In the warmth of the tent and sleeping bag, Oates's rotting skin began to stink.

Though he had hoped he would die during the night, Oates awoke the next morning. Outside, the temperature was again more than forty below zero and a blizzard howled at the men as they huddled in their tent. Oates cursed his life and knew he was only holding the others back. If they had any chance, they had to go on without him. He looked at the others around him, then said, "I'm just going outside and may be some time." He rose and reached for the tent flap. No one tried to stop him.

I turn to look out the window. The leafless, fruitless cherry trees claw at the clouded sky. The restaurant is filled with the smells of people and what they do, their colognes—better on Monday, a little worse each day after—overlaid with stale coffee and hot oil. The ice has melted in my glass, and the Formica beneath it is covered with water rings and bits of salad.

"Dad, I think you're wrong. I think that for reasons neither of us understand, you have gathered up this story from some dark spot inside of you, some place that is so angry about having to face life without Mom that it has created a fiction to make matters even worse."

I'm warming to this now. "Dad, I really believe that this story is something you created to refocus the anger you have about Mom dying. Mother never did any of this. I think it's all inside your head."

My father listens for a while, then takes a long, slow look at me. He sips a little of his water. His eyes fill up with flames. "Well, I remember when *you* were having problems with *your* wife. None of that was 'inside *your* head.'"

More than thirty years before, at a time when I felt like nothing, I had told him about how the woman I loved had sent me on a trip to the East Coast so she could slip away with an orthopedic surgeon. I had needed so badly to touch someone, and my father had accepted that. But now his hatred is so great it has slopped over onto me.

I bristle. My fist tightens around my water glass. I am angry now. I look away, draw a deep breath, and wait.

"The last thing your mother told me was that Michael was not my son. I love your brother, even if I'm not his father. I raised him," he says, tearing up. "He'll always be my son. But I can never forgive your mother."

Michael is my older brother, the oldest child in the family. Because of his age, Michael is the only candidate for bastardization. In truth, Michael doesn't even qualify. Michael was born about three years after my parents married, long after the time frame my father has assigned to my mother's assignation. I've pointed this out to my father on more than one occasion, but he isn't moved by it.

Now I point out to him the Michael has type O blood. I remind my father that he, too, has type O. For a moment, that slows him, but only for a moment. "Wait a minute," he says, "isn't O the most common blood type of all?" Suddenly, in defense of his dementia, my father is a hematologist.

"Dad, I'm telling you, none of this ever happened."

"You don't know your mother like I do. I was there. The last thing she told me, I'm telling you, was that Michael wasn't my son. In the foyer, there at Four Seasons, the day she died. She was sitting there sort of slumped over after lunch. She looked up at me just long enough to tell me. Then she slumped back down. I thought she hadn't eaten, so I wheeled her into the cafeteria there. But they told me she had already had her lunch. So I wheeled her back into the big round place out front and I left her there, where she could find Roy."

Roy, whose name was actually Robert, was a nursing home wreck who, my father imagined, was now my mother's lover. For the last year or so of my mother's life, neither she nor Robert could move from their wheelchairs. The dark spot inside my father's heart was metastasizing.

"Dad, she couldn't speak for days before she died."

"She spoke to me," he states flatly.

When the storm quieted, the final three departed. They took Oates's sleeping bag with them for the first few miles in case they found him alive. At their next camp, though, they gave up hope for Oates and threw out his heavy elk-hide bag to lighten the sledge. But they kept Scott's pile of rocks. Now and then, they added to the pile.

Each day the sun dropped farther below the lidless rim of the frozen Earth. Each night grew longer and colder. For ten days, a vicious wind blew sheets of snow and blistering ice while the men sat in their tent hoping for even a short break in the storm. By now, they estimated they were less than twelve miles from fuel and food, less than a day's walk from surviving. But no break ever came. On March 29, each of the men began writing letters to his family, letters explaining how each would die. The storm continued to beat at the frozen desert that surrounded them.

Where they had thought to find glory, they found instead the tracks and the trash of others, the wind, the ice, and the marble-blue sky.

Nothing mattered now but the letters.

Outside, the snow began to fall more heavily, slamming the tent walls, soldering the seams between tent and ice. Throughout that night and the next, and the next, and the next, the snow fell. So slowly that even the dead didn't notice, the ice swallowed them all.

Ice dreams.

Torrents of ice, buried lakes of ice, continents of ice. The rotted timbers of frozen ships. Ice storms. Shattered masts. Feathered needles piercing flesh, coruscating eyeballs. Frost-blackened fingernails raking at frozen flesh. Emptied sockets, cakes of ice. Screaming sunlight bolting frozen rivets into bare skin. Crystalline horses. Ice-fast sledges. Burning blues and grays. Ice nails driven into a glass-black sky, the nails weeping. Iced women frescoed on ice walls. The frosted faces of the dead and dying. Breath and broken glass. Ice spouts like frozen tornadoes from earth to sky. Shattered pressure ridges, iced floes feeding frenzied seas. The wetness of promise frozen in blood. Ice-white finger bones. And fear, always the wet tongue of fear.

Today, my father tells me he still dreams.

I have a picture of my father and mother when they were about twenty-five, dressed in white cotton pants and shirts, rubber-soled shoes, decked out for boating, maybe. They look young and handsome, they look pleased to be with one another. They look like there is no future, only now. I ask my father if he ever dreams of days like those.

"No," he says. "I dream of arriving in heaven and finding your mother with another man. That nearly kills me. Or," he goes on, "I dream of her mocking me for my stupidity. Good nights, I dream of dying."

"I really miss the ocean," I say to him.

"It was different when it was you. Wasn't it?" he fires back at me. "I remember *you* were pretty upset about it."

My grip slips a little on the glass I am holding, and it overturns. Water and ice spread across the tabletop and run into my lap.

My father continues. "And I've tried but I just can't stop thinking about it. Can't stop being angry about how she screwed me."

I stew silently in my own water. My father picks up the bill from the table, eyes it, and drops a two-dollar tip. "You done?" he asks me.

"Looks like it," I say.

"Let's go then."

My father and I head for his Mercury. I hold his shoulder as he steps into the car—a great white coffin of a car. His shoulder beneath my hand is knobby from too many good-byes. I try to hang on to him, but he drops into the car seat and he is gone.

Two old men in a goddamn Mercury, cruising down the boulevard, wishing for things that can never be. Dried out, closed up, praying for warmer weather.

"You know what the last thing your mother said to me was?" he begins again. I turn on the radio.

November 12, 1912, my father was two months and ten days shy of his first birthday. That same day, a search party spotted the remains of the Scott expedition. As the men in the rescue effort plodded out across the ice, one of them spotted the tip of a tent pole sticking a few inches above the snow. The others gathered, and together they scraped the green ice and crusted snow from the walls of the tent. Inside they found the three dead explorers, the discovery men—Robert Falcon Scott, Edward Adrian Wilson, and Henry Robertson Bowers.

And their letters, too, tales of discovery and sorrow that tailed off into shivering scribbles as the explorers' minds and fingers froze, those letters were in the tent as well. Scott had even penned a note to the public at large, trying to explain the team's and his failure. It was a surprisingly long note full of contradictions and fear, fear for himself and fear for his family. Explanations and obfuscations.

Scott, it seems, was the last to die. He knew how badly he had failed—himself and his men—and as he died, it ate deeply into him. He wrote of all the things that had gone wrong—the weather, the wind, the soft snow, the hard snow, the loss of their ponies so early in the journey, the long hard haul across more than 700 miles of frozen waste, and the wind, the god-awful wind.

Without all of that, people would surely think him a fool to have set out with so little on such a dangerous journey, to have left the safety of England with too little knowledge of where they were going, too little food, too little time, too little strength to conquer such a monstrous adversary as the pole.

He had to tell his story to the world, otherwise no one would understand that it was love that had driven him and his men to this point—love of the Earth, love of adventure—and pride, certainly.

But as he wrote, the ice reached into him and took hold of his mind. His words collapsed. Only fear carried him beyond that point. Scott's final sentence was a plea to Mother England to find it in her heart to care for the dead men's families. England chose to do otherwise.

Beside the tent, as the searchers dug into the ice, they found the explorers' wooden sledge. Once they had chipped away all of the rime and ice, the men could see that the explorers had jettisoned most of their gear to lighten the load as they raced toward food and safety, raced against the creeping cold and the slithering darkness. Extra sleeping hides, tools, equipment—most everything they could have dropped, they did. But they kept a wooden chest filled with more than forty pounds of rocks.

The next March, March of 2003, my father died. By then I had moved him to a full-time care facility. We celebrated his last birthday on a snowy and frigid night in January of that same year. The people who watched over him, my wife and I, and two or three of his housemates all went to a nearby Country Buffet restaurant.

We chose the Country Buffet because a year or so after my mother's death, Dad had taken up with a woman friend, Irene. She had been fond of Country Buffet, and she and Dad had often shared lunch there. Irene later moved to Loveland—not a great distance away, only fifteen miles. But near the end, I had taken Dad's car keys. That year, my birthday card was addressed: "To the Great Car Thief." I told him I would be happy to drive with him to Loveland. But by then Irene had moved to Louisiana where she had family.

In spite of the snow and the cold that evening, my father truly enjoyed himself and never mentioned my mother. Perhaps it was the company, or maybe, in the midst of it all, he had simply forgotten. Outside the parking lot filled up with snow and the snow swallowed everything, even, for a moment, my father's madness.

A week or two later, while Dad was helping to serve lunch, he fell and fractured his pelvis. It wasn't a big fracture. The doctors had to look very closely at the X-rays to see the hairline of darkness that pierced his bones. The fracture was small, but the pain was not. Dad began to spend more time in bed and less time moving around with his walker. He still prayed at night for God to kill him, and slowly, one piece at a time—as only a God would—my father died. On the fourth of March 2003, at about 3:30 in the afternoon, he took his last breath. My sister was with him. I missed his death by about five minutes.

The death certificate claimed my father died from metastatic prostate cancer. His prostate *had* metastasized. Pieces of what had once chosen urine or sperm for him were now in his spine, his ribs, and maybe his mind. But the tumor didn't kill my father. Abandoned by his God, widowed by his wife, dismantled by his son, he just stepped outside for a while and he may be gone for some time.

Signage alteration in Nederland, Colorado

II

On the Lip of Immortality

I don't want to achieve immortality through my work . . .
I want to achieve it through not dying.

—*Woody Allen*

"Dust to dust. Ashes to ashes." To live is to die. Or so
we've been told by teachers and preachers, even by biologists
who should have known better. But as we speak, immortal beings
are roaming this planet—big complex living breathing creatures
not so very different from us. For the rest of us, mortality may
be nothing more than punishment for past sins. Now may be the
time for atonement.

Across the street from where I stand, four men in their late thirties—
dressed only in diapers—carry a coffin bearing a pregnant mannequin.
As I watch, the falling snow turns them pink, then ashen. Suddenly, the
mannequin shudders violently and strains to birth another child. A

DOI: 10.5876/9781607322337.c11

man's voice, tinny through the cheap public address speakers, calls out for an epidural.

Beyond the now-anesthetized mannequin, an emergency vehicle rolls forward. On top of that vehicle stands a pale man with a shock of black hair. He holds what appears to be a flamethrower. Further down the street stands a long line of beetle-like hearses glistening black from the snowfall and snorting smoke as they wait their turn to roll through the streets of Nederland, Colorado.

People in skull masks and skeleton T-shirts dance alongside men and women carrying living people inside coffins. These coffins, though, are only for show. The competition-level racing coffins will come later. And after that, men and women dressed in bikinis will dive through a slot cut in the ice to chill themselves to their very bones in slushy water.

All of this because a few blocks from here, inside a Tuff Shed, there is a dead guy stuffed inside a metal bin full of dry ice, as he has been for the last fifteen years.

Back in Norway, Bredo Morstøl was a fisherman. When he had the time, he hiked and skied as well. Sometimes he liked to paint. In 1967, at age sixty-seven, Bredo retired from his job as director of parks and recreation for the county of Baerum. He and his wife, Anna, wanted more time to enjoy themselves and their grandchildren. And at least one of those grandchildren loved them very much in return. So in 1989 when Bredo died, his grandson Trygve Bauge froze Bredo's corpse on dry ice. It seems Trygve, even as a child, had a thing for cryonics and hoped to open his own institute someday.

Gramps became his first customer, albeit nonpaying.

For a short while, Trygve and his grandfather were at peace. Then Anna died. Depression and fears of nuclear holocaust raised their rat-like heads inside Trygve's brain. Anna's family claimed her body, and Trygve began to look for a safer haven for Bredo. Trygve chose the United States, and a few months later he packed Bredo in fresh dry ice, bought fare for himself and his corpse, and set sail for Oakland, California, and the Trans Time Cryonics facility. Trans Time Cryonics had been "suspending" people since 1974, and Trygve, after some considerable thought, decided Trans Time was a great place to stash Grandpa while Trygev searched for a place to establish his own cryonics institute. The technicians at Trans Time were happy to oblige the pair and quickly settled Bredo at the bottom of a steel bathtub filled with liquid nitrogen.

While Grandpa lay in deep contemplation at −350°F, Trygve set about locating a final resting place for his grandfather and the others Trygve knew were eager to cryogenize themselves while science solved the whole death thing.

In 1993, Trygve settled on Nederland, Colorado, a picturesque little town in the foothills about forty-five miles northwest of Denver. Bredo, once again on dry ice, headed overland for Colorado. Trygve wired his mother, Aud, and told her to meet him and Grandpa in Nederland. Once on site, Trygve began construction of a storm-proof, flood-proof, fire-proof, earthquake-proof, and thermonuclear bomb–proof house. Nearby, he raised a small aluminum shed.

A week later, an unusually heavy package arrived from Oakland, California, and disappeared into the shed. A month or so after that, another equally heavy package arrived from Chicago, and it too quickly fell from sight inside the shed. That box contained the earthly remains of one Al Campbell along with a few hundred pounds of dry ice. Trygve was officially in the cryonics business.

For six months, everything went more or less as planned. But then Trygve got a bee in his bonnet. Trygve was convinced that bathing in ice-cold water was good for one's health. And after a few months of practice, he was also convinced that he could do it for longer than anyone else ever had. So in February 1994, Trygve Bauge invited all the local newspapers and then sat in a 1,500-gallon tank of ice water for one hour and four minutes, breaking the record. Trygve was elated, and the newspapers spread the word about his feat.

That was Trygve's undoing. One of those newspapers just happened to fall into the hands of an INS agent, who duly noted that Trygve's visas had expired. Trygve took it on the lam. But before the end of that year, the INS had Trygve on a plane back to Norway.

That left only Aud behind to tend to matters in Nederland. But once again, because of all the publicity and the search for Trygve, the city of Nederland had discovered that Trygve's nuclear bomb–proof house had neither electricity nor indoor plumbing. That was against the law in Nederland.

The city issued an eviction notice. When a reporter asked Aud about her plans for the future, she explained that she could not leave the premises or Grandfather and Al would thaw out. The reporter, in a state of shock, sought out the chief of police, certain he would have something interesting to say about all of this.

Sirens blaring, the police arrived at Aud's a few moments later. Threats were issued, and Aud's arrest seemed inevitable. But it turned out that Nederland had no law that prevented the in-home storage of human remains—frozen or otherwise.

That evening, the mayor called an emergency town meeting, and the good citizens of Nederland passed a new law. Henceforth it would be illegal to store the "whole or any part of the person, body, or carcass of a human being or animal or other biological species which is not alive on one's property."

But no one was eager to haul Bredo and Al from their frozen sleep or to face Aud with the news that all was lost and Grandpa's frozen interlude was about to be terminated. To obviate this angst, the frozen remains of Grandpa Bredo were "grandfathered" in, and Bredo and Al remained in their shed. But no sooner was one problem solved tham another arose. About this time, the INS took after Aud. Soon, she too was back in Norway. For his part, Al Campbell—at his family's behest—made the return trip to Chicago. But Grandpa remained. Creating a bit of a conundrum. Into the crevasse leapt Trygve.

To forestall the unthinkable, he hired Bo Shaffer, the CEO of Delta Tech—a nearby environmental company—to take care of Grandpa. Once a month, Bo was to haul dry ice from Commerce City (near Denver) to Nederland and repack Grandpa's bearings. Because of that, Bo came to be known as the Ice Man. And for fifteen years, the Iceman has cometh, and Bredo Morstøl has been lying in a state of suspended (if not ended) animation. Bo has since founded the International Cryonics Institute and Center for Life Extension (ICICLE) dedicated to research on life extension.

In 1999, another tragedy struck. A mighty wind blew through Nederland, and among the casualties was Grandpa's shed. Volunteers quickly threw up pieces of plywood and banged in tenpenny nails. But it was clear these efforts wouldn't hold Bredo and his frozen aluminum box for long.

To the rescue rode the folks at Tuff Shed with new accommodations for the stiff and its icy cocoon. Every March since then, the town of Nederland has celebrated "Frozen Dead Guy Days," honoring Bredo and his icy past.

"As caretaker," the Ice Man told me, "I've seen the gamut of reactions to Grandpa—some people are truly fearful; some are just amazed; some are highly amused; and some are highly offended, for one reason or another."

Events include coffin races, parades of hearses, costumes, considerable tom-foolery, and, for those willing, a chilling dip beneath the ice. All because of a single dead guy.

Well, maybe not all.

Surely, many of those who come for Frozen Dead Guy Days come for something other than the dead guy. Though their reasons, I'm sure, vary, standing amid all of that revelry, all the skulls, all the black gowns, I feel certain that many come just to thumb their noses at death. But after speaking with Bo, I've changed my mind.

"Facing death in an 'everyday' situation like [this] festival," Bo says, "allows people to contemplate death in a safer environment—one where they are (mostly) certain they will *not* die—and perhaps become more at ease with their own inevitable demise."

A few others, I suspect, come to see just what can be done with a Tuff Shed and a load of dry ice. Death, after all, is in our veins.

Last year, attendance exceeded 5,000.

And every year, at the beginning of November, hundreds of thousands of people throughout North, Central, and South America don costumes, eat sugar skulls, drink, and croon to mariachi guitars to celebrate Los Dias de los Muertos (the Days of the Dead)—a tradition, with mestizo roots, that honors the dead. It seems likely that many of these celebrants come for the same opportunity to consider the visage of the Grim Reaper under semisafe (sometimes semiconscious) circumstances.

Because, for all of us, mortality is a heavy cross to bear.

But our teachers and our preachers have told us that death is our lot in life. In fact, they have told us that death is every living thing's lot.

In truth, we've been lied to.

IMMORTAL BEINGS PAST AND PRESENT

According to paleontologists, life on Earth began about three billion years ago. No doubt, first life was simple. Still, those first creatures must have eaten and drunk, grown and reproduced, as all living things do. But just how they did all of that remains a mystery, because at the outset, and for about the next

750 million years, life was not only simple but squishy—no bones, no shells, no exoskeletons.

The fossil record doesn't offer us much insight into just what early life looked like. But in the rare cases where bacteria did leave fossils—some cyanobacteria or bacteria in fossil amber—those billions-of-years-old bacteria looked a lot like twenty-first-century bacteria. So it seems reasonable to assume we can use what we know about modern bacteria to speculate about how first organisms looked and behaved.

Modern bacteria live by the simplest of means. They are little more than a bit of DNA surrounded by a lipid membrane and a cell wall. But in spite of their simplicity, bacteria are the most successful of all living beings. Because of bacteria, our skies and oceans are blue, and our atmosphere is rich with oxygen. If you could add up the numbers of all other living things—fungi and insects included—your total would come nowhere near the number of bacteria. As Stephen J. Gould once said, "This is not the age of man. It is now and always has been the age of bacteria."

Bacteria reproduce by cell division—one cell becomes two, then four, then eight, then sixteen, and so on. There is no such thing as an old bacterium. Though there is now some evidence that bacterial divisions do not always produce equal progeny, most do. So even after thousands of divisions, most bacteria retain all of their youthful vigor.

By implication, that means that for the first several hundred million years of life on this Earth, death from natural causes was unheard of. Living things simply did not grow old and die. Of course, they got eaten and infected, squished, left out to dry, and a lot of other things, but age never fazed them.

Flash-forward three billion years or so. Life has diversified beyond the imagination of the living. From kangaroos to duck-billed platypi, from brown recluse spiders to mantises, from fish to fungi, life seems to know no bounds. But among all of those living things, bacteria still rule supreme.

Far more than 90 percent of all living things on this planet are and always have been bacteria. And that means that far more than 90 percent of the living things on this planet are and always have been immortal. You can, of course, kill bacteria, but under normal circumstances, they will not kill themselves, as our bodies do.

Immortality is not some pie-in-the-sky pipedream. It is the way of most life on this Earth. Bacteria have it made. The rest of us, like Grandpa Bredo, got a raw deal.

Sure, you say, but that's bacteria, and bacteria are, well, bacteria. But there's archaea, too. Archaea are single-celled lesser-known cousins of bacteria, and they are everywhere, including inside each of us. They, too, are immortal. So are some yeasts and other fungi.

But none of these organisms seems very much like a human being. So maybe agelessness is just a feature of strange and primitive life forms. Still, bacteria are a lot more like human cells than they are like rocks, suggesting age is not an inherent property of life. And second, there's Pando.

Pando lives on a hillside inside of Fish Lake National Forest, a sizable stretch of land in east central Utah. He's big—over 4,700 acres—and heavy—in excess of 13 million pounds. But what really gets most people about Pando isn't his size or his weight, it's his age. Minimally, Pando is over 80,000 years old. More likely, Pando is well over a million years old. Pando is old enough to have seen the constellations change shape, the Earth's magnetic poles reverse, and the glaciers advance and retreat and advance once more.

Fish Lake National Forest and Pando are a little off the beaten track, in between the towns of Loa and Salina, Utah. Because of that, relatively few human beings have had the opportunity to meet Pando personally. But most all of us are on speaking terms with one of Pando's relatives. Pando is a male aspen tree—a complex, eukaryotic, multicellular organism, just like we are. However, unlike us, Pando is, for all intents and purposes, immortal and subject to death only from accident. No old age, no natural death.

As is the case with Pando, a grove of aspen trees is often a single living entity, linked together underground, all of the apparently individual trees simply arms of one creature. The death of one aspen stem represents no more than we might experience when we trim away our nails or the dead skin on our feet. A stem dies, Pando continues.

Aspen trees don't make deals with the devil. In spite of what we think we know about life and the inevitability of death, aspen trees simply persist. Clearly, nothing in Pando's biology leads inescapably to death, as we have been told all biology must.

In one very real sense, a grove of aspen trees is a clone of genetically identical elements, each connected to the other. Many other plants have also adopted this path in life. And many of them, it turns out, also live very long lives—from a Mediterranean sea-grass colony that is 100,000 years old to a Pennsylvania huckleberry bush at 13,000 years old, to a colony of 9,550-year-old Norway spruce in Sweden.

Old clones, for certain. Maybe long life is the sole property of clones. But it isn't. Many individual plants—plants a bit more like us than aspens are—also live inordinately long lives. In 1964, a Great Basin bristlecone pine by the name of Prometheus stood gracefully upon the flank of Wheeler Peak, one of the peaks in the Snake Range inside of what is now Great Basin National Park in eastern Nevada. Those who knew of Prometheus knew that this tree was very old. Whether they knew that Prometheus was older than any of the surrounding trees remains unclear.

Also in 1964, Donald Curry was a graduate student studying climate dynamics at the University of North Carolina in Chapel Hill. Curry's main interest was in investigating climate change using dendrochronology—tree-ring analysis. The more rings you could obtain from a single tree, the more you learn about past climate change. Somehow—word of mouth, newspapers, *National Geographic*, television—Curry found out that some of the oldest living trees were bristlecone pines in Nevada. In the summer of that same year, Curry met Prometheus.

Several times, Curry had attempted to obtain continuous tree-ring samples in cores taken from Prometheus, but with little success. Because of that, somewhere along the way, Curry had either asked for or simply been granted permission to cut down trees, including Prometheus.

On that particular day, Curry may have been having trouble with his corers and broken one of the only two he carried. From his point of view, he was out of options. In the summer of 1964, Curry cut Prometheus down.

What Prometheus's severed trunk revealed shook even Curry. On the day of his death, Prometheus was likely over 4,862 years old. The oldest nonclonal organism ever found. Later, more accurate estimates place Prometheus's age at no less than 5,000 years.

Long before Christ and his apostles pulled fish from the waters of Galilee, before the Norman invasion, before the Buddha sat beneath the Bodhi Tree, before Muhammad sought out his cave, Prometheus watched sunrises and sunsets from a mountaintop in Nevada. Who knows how long Prometheus might have gone on doing just that if he had not met Donald Curry.

Fortinggall Yew, a tree in a churchyard in Perthshire, Scotland, is somewhere between 2,000 and 5,000 years old.

Very long-lived, nonclonal plants. The more you look the more you find. Many creatures on this planet live far longer lives than we do, and they all aren't aspen trees or huckleberry bushes.

We tend to think of time and age in terms of human life spans, in packets of 100 years or so. Everyone, or nearly every one of us, realizes that dogs and cats, butterflies and moths, grasshoppers, insects, geese, and house flies don't live as long as we do. Their lives evaporate in less than a century, or so we notice. But the spruce tree behind my house? I have no idea how old it might be. It's been there for as long as I can remember, and I fully expect it to be there after I am gone. So it fades rapidly from my catalog of the living. For many of us, trying to put an exact age on something like that spruce seems a waste of time. So I forget about it, just as I forget about the Rocky Mountains looming to the west, since, from my perspective, they have always been there and always will—so old already that at times they nearly disappear.

But that doesn't mean that they are not old beyond my imagining. The same can be said of the living.

One of the neatest tricks in all of the animal kingdom belongs to *Turritopsis nutricula*. These beautiful, small jellyfish have hitched rides in the ballast water of ships heading to ports all over the world. Today, they swarm in warm tropical seas everywhere. But their facility at thumbing down rides isn't what sets them apart.

T. nutricula can rejuvenate themselves in the truest meaning of that word. Like many other jellyfish, *T. nutricula* begin life as eggs that hatch into small polyps, then mature to the sexually competent adult, and reproduce. Where *T. nutricula* differ is that instead of dying after reproducing, as other jellyfish do, these jellyfish revert to the polyp stage, and the whole process begins again. *T. nutricula* become adults long enough to reproduce, and then revert to teenagers with all of their original vitality and longevity. Immortal animals.

Fish, like koi, may live for more than 200 years. At least one quahog clam has been found that was between 400 and 410 years old. Lobsters can live for more than 140 years, and turtles may survive as long as 250 years.

Fish and clams, lobsters and turtles—but what about mammals, those animals most like us?

One summer about ten years ago, Craig George was helping Alaskan native and whale hunter Billy Adams to dismantle his kill. George noticed a large scar in the skin of the whale's head and pushed his knife in to investigate. The blade struck stone. George dug deeper and extracted a stone harpoon point. No native hunter had used a harpoon point like that in well over 100 years. Obviously, this whale had reached maturity more than 100 years ago. By 2001, more than six such points had been recovered from freshly killed bowhead whales.

George got hold of Jeffery Bada, then working at the Scripps Institution of Oceanography in La Jolla, California. Bada had worked out a way to determine a whale's age using the level of aspartic acid in the lenses of the whale's eyes. Bada was very interested in George's findings. So George plucked the eyeballs from several dead bowhead whales—some infants, some adults. He retrieved forty-eight eyeballs in all, each about the size and shape of a billiard ball. Then he froze them and shipped them off to Bada.

Bada worked his way through all of them. Five males among those forty-eight registered ages of 91, 135, 159, 172, and 211 years. Investigations in populations of blue and fin whales make it clear that other aquatic mammals also live very long lives.

These are air-breathing, lactating, baby-nursing, big-brained, and highly intelligent mammals that—even without the miracle of modern medicine and seat belts—live at least twice as long as human beings.

When it came to picking the genetic straws, did we simply end up with the short one?

Maybe.

DEATH IN OUR GENES?

What happened? Why, in the face of so much longevity—even immortality—do we humans continually suffer the indignity of death within our 100 years? Maybe it is in our genes.

You can make long and relatively convincing arguments that large, fairly rapidly reproducing omnivores like us put large demands on our environments. If we regularly lived hundreds of years beyond our reproductive ages or worse, reproduced for hundreds of years, in a very short time we would completely deplete all of our resources (something we've done a pretty good job of even with relatively short life spans) and vanish as a species. So maybe human death arose as a biological adaptation, a few mutations to prevent us from, or at least slow the process of, destroying ourselves.

There are arguments against this, based on the idea that among populations where predation is the leading cause of death, there is no need for a genetic limitation to aging. In other words, if an impala is most likely to die in the jaws of a lion, impalas don't need genes to force them to hang up their running shoes. Still, animals that are most likely to end up as some other animal's lunch all show signs of growing older. And those changes make it even more likely that an individual rabbit or mouse or gazelle will end up on someone else's dinner plate. So even among these "food" animals, genes—even if only indirectly—appear to shorten lives.

But no matter which side of the argument you come down on, one thing is clear. Most every cell inside a human body has a built-in self-destruct mechanism. When any of these cells reaches a certain age, the time bomb detonates, and the cell dies. And for each cell, the roots of this self-termination program are written in its genes. That seems like a protective mechanism of some sort, and no bacterium has anything like it.

But protective or not, for the first time, our genetic heritage and the sins of our fathers no longer matter.

Reanimating the Dead

The coffin races have ended. The hearses pass by in somber black, and the crowd seems a little less jubilant than it was an hour or so ago. It is snowing harder. Surely the snowfall and the dropping temperatures are part of the crowd's darkening mood, but so too is death.

The ice-swimming contest is still to come; maybe that will lift people's spirits once again. If not, perhaps the wine and beer will do it. Up the street, Bredo

Morstøl continues his lonely vigil, hoping for a change in the weather. My wife and I wander off to find something warm to drink. The masked revelers look around once as the dim shadow of a raven passes over them, then they too begin to search for other diversions. Maybe the Grandpa-Morstøl-look-alike contest will hold them for a while. Mortality is, indeed, a heavy cross to bear.

But what all of these old plants and animals tell us most clearly is that the weight of that cross is not shared by all living things equally. Death is not an essential element of life. Very long-lived and even large immortal beings do just fine on this Earth. That means that what stands between us and immortality isn't some unbreakable law of nature. We grow old and die because of nothing more than our profligate past and our genetic heritage.

At the beginning of the twenty-first century, we can change our genes almost as easily as we change our jeans. If we wish to reach for the golden ring of old age, perhaps even immortality, all we need do is figure out which of our genes are killing us.

Epilogue

Leonardo, like most of his fellow men of science, believed mind and emotions throbbed inside the human heart. Vesalius maintained that human emotions and human intellect sparked inside human brains. Neither gave much thought to human thymuses, and neither imagined all the things they could not see or touch or hear or taste or smell beneath the stench of the preservatives.

Regardless, each infused human bodies with human art and instilled in all of those to come a sense of something more than just the art, more than just the science. Even working with a small scientific pallet, they knew that the human whole was greater than the sum of human parts. That was their art.

Vesalius gave vision and the optic chiasm to his brain. And even though he had no idea what happened inside the optic chiasm, his art clearly shows he suspected something twisted. Leonardo knew nothing about things like the S/A and A/V nodes that cause our wrists to throb and allow human fingers to probe inside of human chests. But to look at their art, it seems that lack of knowledge proved only a small

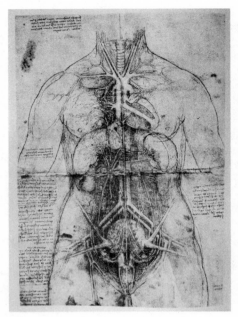

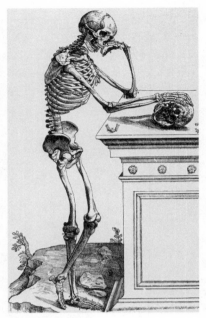

Leonardo's center of being   *Vesalius's thinking man*

obstacle. Each anatomist was there in the science and the art. Leonardo's hearts held more than just atria and ventricles; they held human hope and Leonardo's eyes. Vesalius's skulls held more than human brains; they signaled an end to darkness and they held crystalline visions of the future.

Ultimately, it seems to me, it isn't as much about the absolute completeness of the science as it is about understanding what it means to be a human being in this world at this moment. In spite of their handicaps, Leonardo and Vesalius captured that.

I feel certain we are still nearly as ignorant of our surroundings and our potentials and our biology as both of these anatomists were almost 600 years ago. There is so much that is still, and perhaps forever, beyond us—the light we see by, the very slow, the very small, the very large, our selves. But, just as it did with Leonardo and Vesalius, that puts only very small limits on us. Our greatest limits are self-imposed.

Artists and scientists seek the same knowledge, seek to attain the same grasp on an ungraspable universe. Poets ask the same questions as brain surgeons. Physicists perform the same experiments as playwrights. We all seek to know the world through human eyes—clearly an impossible task. Seemingly, even more foolishly, we seek within our selves for knowledge of our selves.

The saying goes something like, "We are limited only by our own imaginations." Sort of, that's true. But our ignorance limits us as well. And not just ignorance of the things we don't know or can never know, but the self-imposed ignorance of things we refuse to see.

Perhaps if we try using the right eye of science and the left eye of art.